THE ENERGY FACTBOOK

THE ENERGY FACTBOOK

RICHARD C. DORF
University of California

An Energy Learning Systems Book
McGRAW-HILL BOOK COMPANY
New York, St. Louis, San Francisco, Auckland, Bogotá, Hamburg, Johannesburg, London, Madrid, Mexico, Montreal, New Delhi, Panama, Paris, São Paulo, Singapore, Sydney, Tokyo, and Toronto

Library of Congress Cataloging in Publication Data

Dorf, Richard C
The energy factbook.

(An Energy learning systems book)
1. Power resources I. Title. II. Series: Energy learning systems book.
TJ163.2.D66 333.79 80-18736
ISBN 0-07-017623-X (hbk.)
ISBN 0-07-017629-9 (pbk.)

1234567890 KPKP 89876543210

CONTENTS

PREFACE

The energy crisis is of concern to us all, yet only partially understood. The objective of this book is to provide information about the uses of energy, supplies of fossil fuels, alternative energy sources, and various policy alternatives for the U.S. and the world.

The purpose of this book is to provide facts and information about energy and therefore contribute to comprehension and improved judgement. The book is intended for anyone who needs to make decisions that in part require information about energy supplies and uses. Therefore, the book is to be used as a desk-top item for the manager, industrialist, engineer, planner, governmental analyst, and other professionals.

This book has been developed with the assistance of many individuals to whom I owe a debt of gratitude. I wish to acknowledge Ms. Suzy Woolf who typed the manuscript and Mr. Robert Entwistle who inspired the author. Finally, I dedicate this book to a dedicated student and inspirational daughter, Christine Joy Dorf.

Richard C. Dorf
Davis, California
April, 1980

THE ENERGY FACTBOOK

1

ENERGY AND SOCIETY

The operation of our technological society depends upon the production and use of large amounts of energy. Many of the world's present problems are closely related to problems of energy distribution, dwindling fossil-fuel supplies, and environmental effects of various methods of energy production and usage. Energy is not only a commodity, it is also an idea—an intellectual concept which stands out in the history of modern scientific and engineering thought.

The problem of energy use and availability is common, to a greater or lesser extent, throughout the whole world. While the industrialized nations of Europe and North America depend heavily upon fossil fuels for their industrial processes, the developing nations also desire to increase their technological capabilities and thus their use of energy in its various forms. In addition, due to the unequal distribution of fossil-fuels throughout the world, profound economic and political issues are associated with energy use.

Energy, the ability to accomplish physical work, is featured mainly as an input of economic processes and as an intermediate good. It is rare to find energy demanded for its own sake. Energy is usually valued as an input in some process of production or utilization which results in a final product.

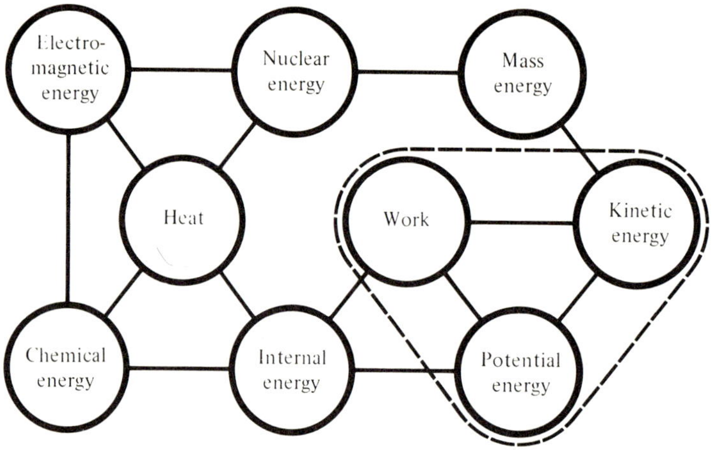

Forms of energy and their interconnections. Potential energy can be converted to kinetic energy, for example. The dashed lines incorporate the forms of energy relevant to macroscopic mechanics or the mechanics of large masses as contrasted to particles.

CONVERSION OF FORMS OF ENERGY

From *To*	*Mechanical*	*Electrical*	*Heat*	*Chemical*	*Nuclear*
Mechanical	Gears, Piston	Microphone, Electric Generator	Friction		
Electrical	Loudspeaker, Electric Relay		Electric Heater	Electrolysis, Battery	Particle Accelerator
Heat	Turbine	Thermo-couple	Heat Exchanger		
Chemical	Rocket, Combustion Engines	Battery, Fuelcell	Fires, Boilers	Chemical Processes	
Nuclear			Nuclear Reactor		
Solar		Conversion Cells	Collector	Photosynthesis	

FORMS OF ENERGY

Kinetic	Potential
Electrical	Magnetic
Heat	Chemical
Nuclear	Sound
Light	Mass

Energy is a property of matter and takes many forms.

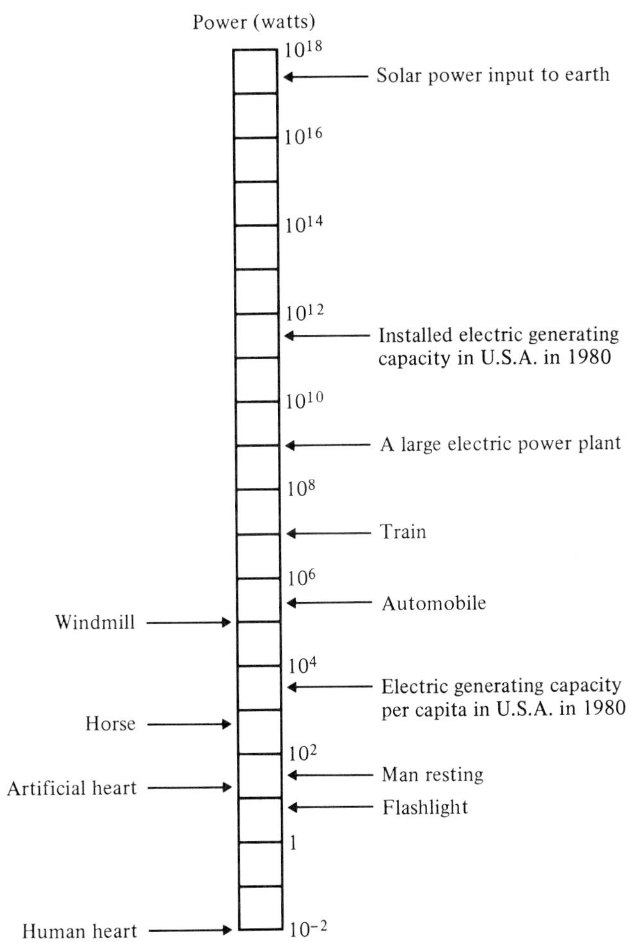

The power of various devices and of animals and humans in watts. The *power* of a device is the time rate of delivery of energy from the device. The unit of power in the SI system is joules/second or watts.

THE INTERNATIONAL SYSTEM OF UNITS (SI)

Basic Units	*Unit*	*Symbol*
Length	Meter	m
Mass	Kilogram	kg
Time	Second	s
Temperature	Kelvin	K
Electric Current	Ampere	A
Derived Units		
Velocity	Meter per Second	m/s
Area	Square Meter	m^2
Force	Newton	$N = kgm/s^2$
Energy	Joule	J = Nm
Power	Watt	W = J/s

The unit of energy is joules and the unit of power is watts in SI.

ENERGY CONVERSION FACTORS AND DEFINITIONS

Unit	*Definition*	*Multiply By*	*To Obtain*
Kilowatt-Hour	One Thousand Watts for One Hour	3.6×10^6	Joules
British Thermal Unit	Heat Needed to Raise the Temperature of Water by 1 Degree Farenheit	1055	Joules
Therm	100,000 Btu	29.3 1.055×10^8	Kilowatt Hour Joules
Calorie	Heat Needed to Raise the Temperature of Water 1 Degree Centigrade	4.184	Joules
Calorie = One Kilocalorie	1000 Calories (Used in Dietetics)	4184	Joules
Electron Volt	One Electron Passing Through One Volt	1.602×10^{-19}	Joules

LAWS OF THERMODYNAMICS

1. *First Law: Conservation of Energy*

The law of the conservation of energy states that: the total quantity of energy in the universe is constant. In other words, energy can neither be created or destroyed. The study of changes of energy from one form to another and the transport of energy from one place to another is called "thermodynamics" (from Greek words meaning heat motion). For that reason the law of the conservation of energy is often called the *First Law of Thermodynamics.*

2. *Second Law: Entropy*

The *Second Law of Thermodynamics* may be stated: every system left to itself changes in such a way as to approach a definite final state of rest or equilibrium. The diffusion of materials from regions of high energy concentration to low energy concentrations; the passage of heat from hot to cold bodies are illustrative of the process.

Many scientists have defined entropy as the practical measure of the random element or chaos which can increase but never decrease in the universe. This means the Second Law of Thermodynamics can be stated: the entropy of an isolated system increases or remains the same. Entropy is a measure of the extent of disorder in a system or of the system. Entropy is said to always increase with time. For an engine or heat device it may be stated that any device operating in a repetitive cycle heat flow out of one part of the system during one cycle cannot be transformed wholly into mechanical energy (via work) but must be accompanied by heat flow into a cooler part of the system. In brief, heat cannot be completely transformed to work by any cyclical machine.

PREFIXES FOR POWERS OF TEN

Power	*Prefix*	*Symbol*
10^{12}	Tera	T
10^{9}	Giga	G
10^{6}	Mega	M
10^{3}	Kilo	k
10^{-2}	Centi	c
10^{-3}	Milli	m

CONVERSION FACTORS FOR CONVERTING TO SI UNITS

From	*Multiply By*	*To Obtain*
Length		
Inches	25.4	Millimeters
Feet	30.48	Centimeters
Miles	1.609	Kilometers
Area		
Square Feet	0.09290	Square Meters
Square Miles	2.590	Square Kilometers
Acres	0.4047	Hectares (Equiv. to 10^4 Sq. Meters)
Volume		
Gallons	3.785	Liters
Cubic Feet	0.02832	Cubic Meters
Barrels (U.S. Petroleum)	0.1590	Cubic Meters
Speed		
Miles per Hour	0.4470	Meters per Second
Mass		
Pounds	0.4536	Kilograms
Ton (2000 lbs.)	0.9072	Metre Tons (1000 kg, also called Tonne)
Force		
Pounds - Force	4.448	Newtons
Work Energy Power		
Foot-Pounds-Force	1.356	Joules
British Thermal Unit	1055	Joules
Horsepower	746	Watts
Calorie	4.184	Joules

The flow diagram of energy use on an illustrative farm. The circles are symbols for energy sources. The symbol appearing as a tank with nutrients is a storage symbol. The flows are given for one day. The farm saleable output is 2×10^8 joules = 47,800 kilocalories. The external inputs are the solar energy. Nutrients stored in the soil also provide energy to the food production process. As a result of the production process, energy is degraded to heat and waste. The farm product is then equal to 2×10^8 joules or 47,800 kilocalories. Since the average person consumes about 3000 kilocalories per day, this farm can supply food for about 16 persons. *Source:* H. T. Odum, Environment, Power and Society, Wiley and Sons, Inc., New York, 1971.

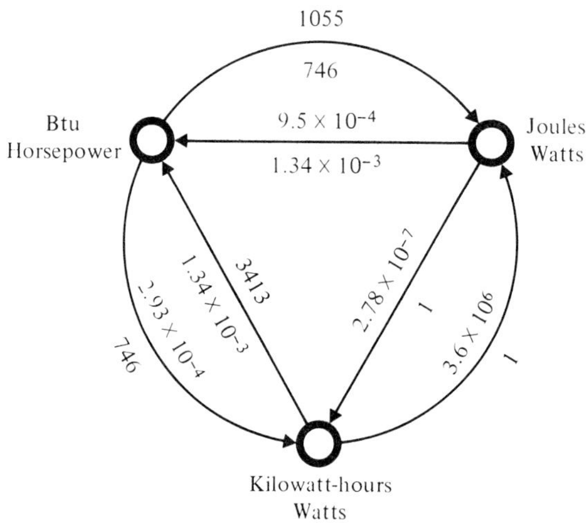

A triangle of energy units conversion. The number given above the line converts the energy units and the number below the line converts the power units.

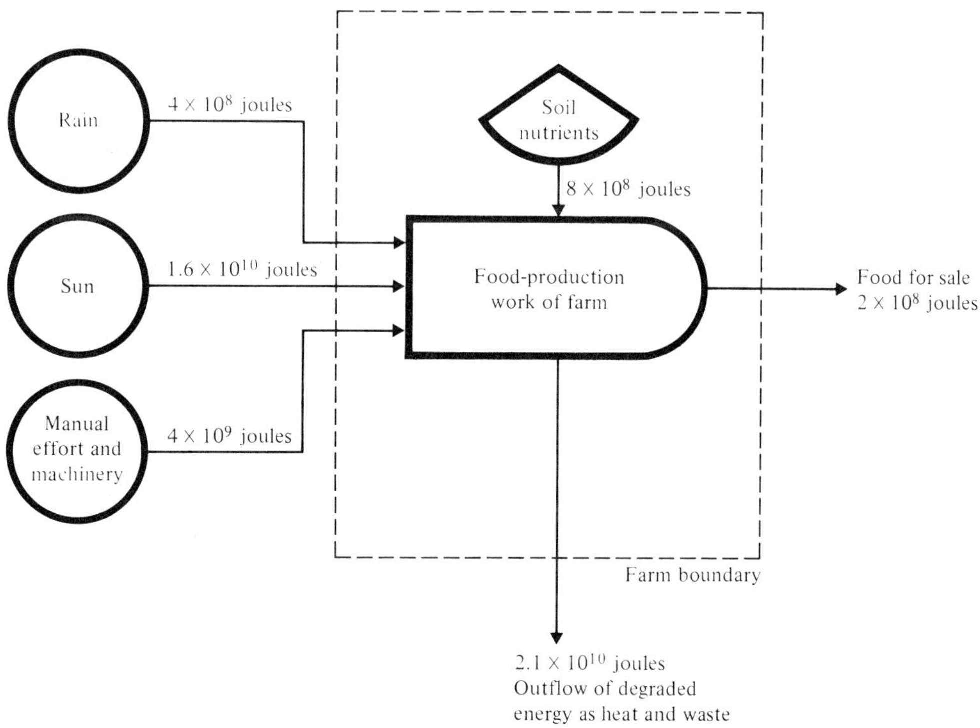

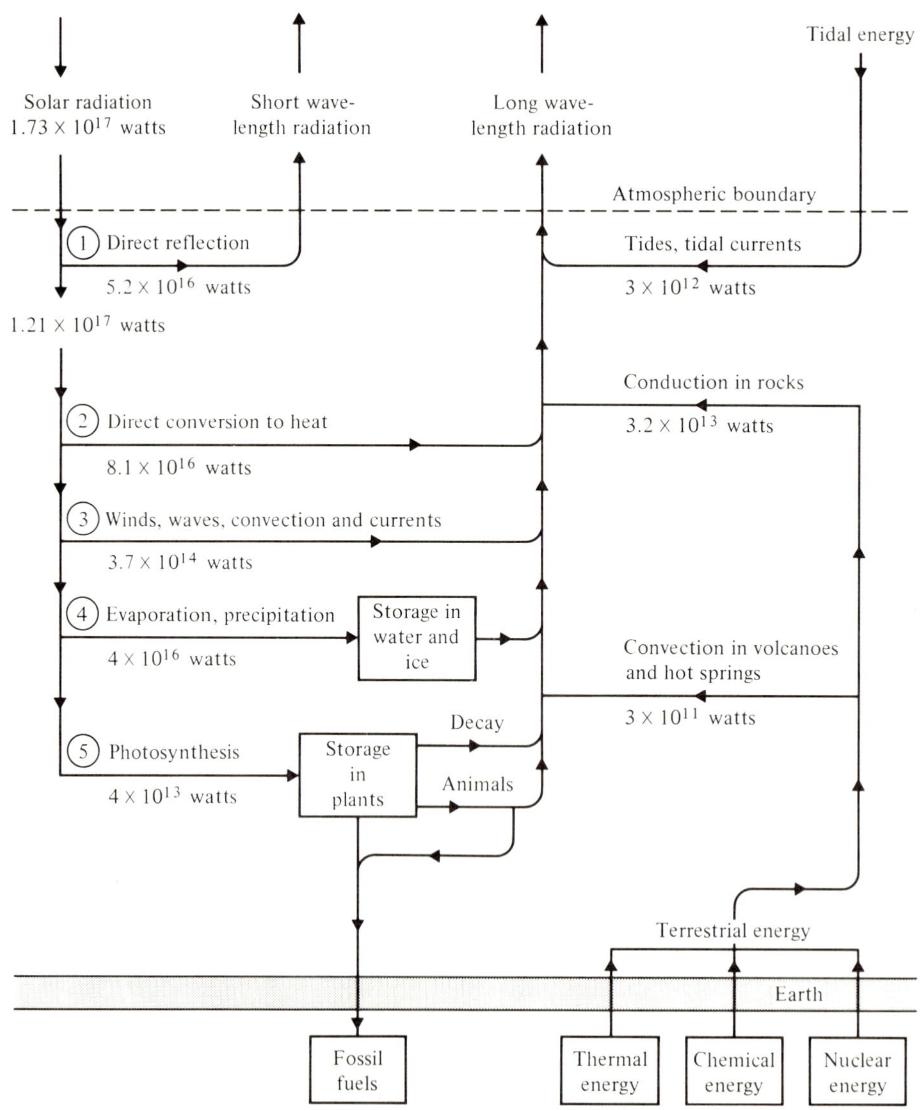
Solar radiation
1.73 × 10^17 watts
Short wave-length radiation
Long wave-length radiation
Tidal energy
Atmospheric boundary
1 Direct reflection
5.2 × 10^16 watts
Tides, tidal currents
3 × 10^12 watts
1.21 × 10^17 watts
Conduction in rocks
3.2 × 10^13 watts
2 Direct conversion to heat
8.1 × 10^16 watts
3 Winds, waves, convection and currents
3.7 × 10^14 watts
4 Evaporation, precipitation
4 × 10^16 watts
Storage in water and ice
Convection in volcanoes and hot springs
3 × 10^11 watts
Decay
5 Photosynthesis
4 × 10^13 watts
Storage in plants
Animals
Terrestrial energy
Earth
Fossil fuels
Thermal energy
Chemical energy
Nuclear energy

Energy flow diagram for the earth. The rate of energy flow from the sun can be calculated to be equal to 1.73×10^{17} watts. The tidal energy input is 3×10^{12} watts. Therefore, the sun supplies essentially all of the energy flowing to the earth. Terrestial energy escaping from the earth is about 3.2×10^{13} watts, ten times larger than the tidal energy.

When the solar power enters the earth, a large part of it, 30% is reradiated or reflected back into space as short wave length radiation (flow 1). The direct conversion of part of the remaining energy is reradiated as long wave length radiation in flow 2. Another part, in flow 3 sets up differences of temperature in the atmosphere and oceans in such a manner that the connective currents produce the winds, ocean currents, and waves. This mechanical energy is eventually dissipated into heat and radiated into space.

Still another part, flow 4, follows the evaporation, precipitation and surface runoff channel of the hydrologic cycle. Heat energy is absorbed during the evaporation of water, but it is again released when the water is precipitated. However, the water vapor, being a part of the atmosphere, is convected to high elevations by means of the convective energy already discussed; and, when precipitation occurs at these elevations, the water possesses potential energy, which again is dissipated back to low-temperature heat on the descent to sea level. It is this energy that is responsible for all precipitation on the land, and for the potential and kinetic energy of surface lakes and streams.

A final fraction of the incident solar energy is captured by the leaves of plants by the process of photosynthesis. In this process, solar energy is stored as chemical energy. If energy could be stored in plants and retained, as in firewood, indefinitely the aggregate amount of stored energy could increase rapidly. In nature, however, decay occurs and the release of stored energy is nearly equal to the rate of photosynthesis.

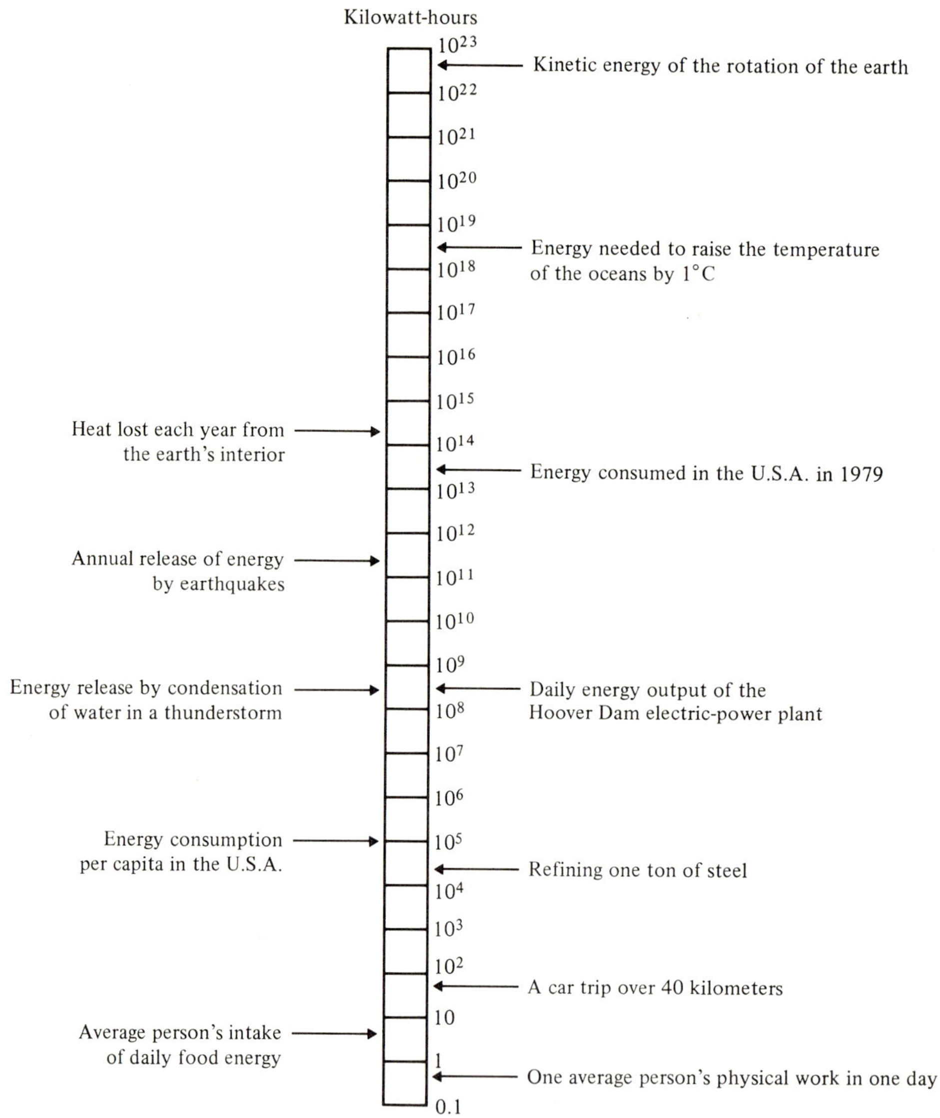

Typical Energy Expenditure in kilowatt-hours. (One Kilowatt-hour = 3.6 megajoules.)

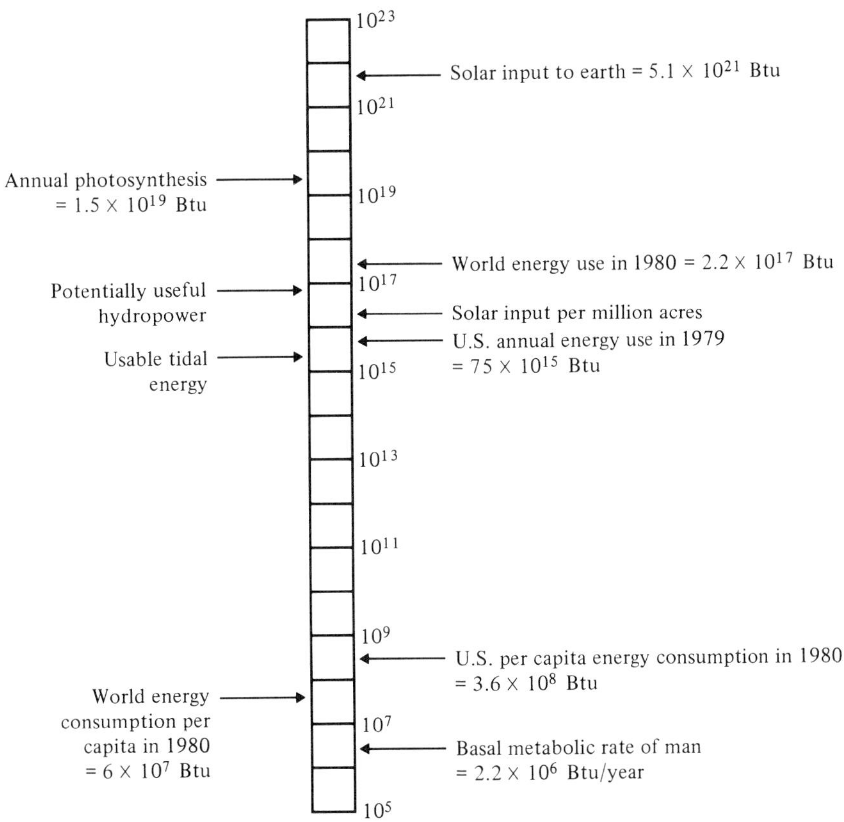

The solar input to the earth and related energy rates. The unit of energy is the British Thermal Unit. Ten thousand Btu's equals 2.93 kilowatt-hours. Most of the energy utilized by mankind until the advent of nuclear power originated in the sun. The solar energy is absorbed and stored in plants and provides the energy in foodstuffs. The energy of coal and oil is also derived from sunlight. Energy from the sun also gives rise to the winds in our atmosphere and the rain that fills our rivers and drives waterwheels.

2
LAWS OF THERMODYNAMICS

The artful manipulation of energy has been an essential component of human ability to survive and to develop socially. The use of energy has been important in the development of a supply of fuel to provide physical comfort and to improve the quality of life beyond the rudiments of survival. The utilization of energy depends upon the availability of resources and development of the technological skill to use them.

It is only in recent times that we have begun to fully utilize the resource of fossil fuels. Previously, people depended upon solar heating, draft animals, and energy. The development of power devices to convert energy into useful work has been a recent historical development.

The economy of the United States is highly energy-intensive and per capita use continues to grow. During the period 1945 to 1965, energy consumption increased at an annual rate of 3%. During the period 1965 to 1971 the rate accelerated to 4.8%. Then the rate slowed again to 3% following the increases in gasoline and heating oil prices in late 1973 and subsequently.

World energy consumption has also grown rapidly during the period 1950 to the present. During this period world energy consumption increased at an annual rate of 4.5%.

ENERGY CONSUMPTION IN THE UNITED STATES

Year	*Total Energy Consumed (10^{12} Btu)*	*Population (Thousands)*	*Index of Industralized Production*	*Total Energy per Capita (Millions Btu)*
1850	2,500	24,000	—	105
1900	8,300	75,000	—	110
1920	19,782	106,466	16.6	186
1925	20,809	115,832	19.9	180
1930	22,288	123,077	20.2	181
1935	19,107	127,250	19.4	150
1940	23,908	131,954	27.8	181
1945	31,541	132,481	44.6	238
1950	34,153	151,241	47.4	226
1955	39,956	164,309	61.1	243
1960	44,816	179,992	68.8	249
1965	53,969	193,815	90.7	278
1970	67,444	203,185	106.7	330
1972	72,108	206,073	106.4	345
1973	75,561	210,400	—	359
1974	73,941	213,500	—	346
1976	74,413	215,500	—	345
1978	77,672	217,785	—	357
1979	79,114	218,850	—	361

The British Thermal Unit (Btu), a common unit of energy, is the heat needed to raise the temperature of 1 lb. of water by 1 degree Fahrenheit. One Btu equals 1,055 joules.

ENERGY CONSUMPTION PER CAPITA IN SELECTED COUNTRIES IN 1968 AND 1971

Country	*1968*	*1971*	*Increase (%)*
USA	312	337	8.1
France	98	117	19.7
W. Germany	134.5	156.5	16.4
Great Britain	148.7	165	11.0
USSR	121.4	136	12.0
Japan	75.5	98	29.7
World	52	58	11.1

All per capita figures in millions of BTU's.

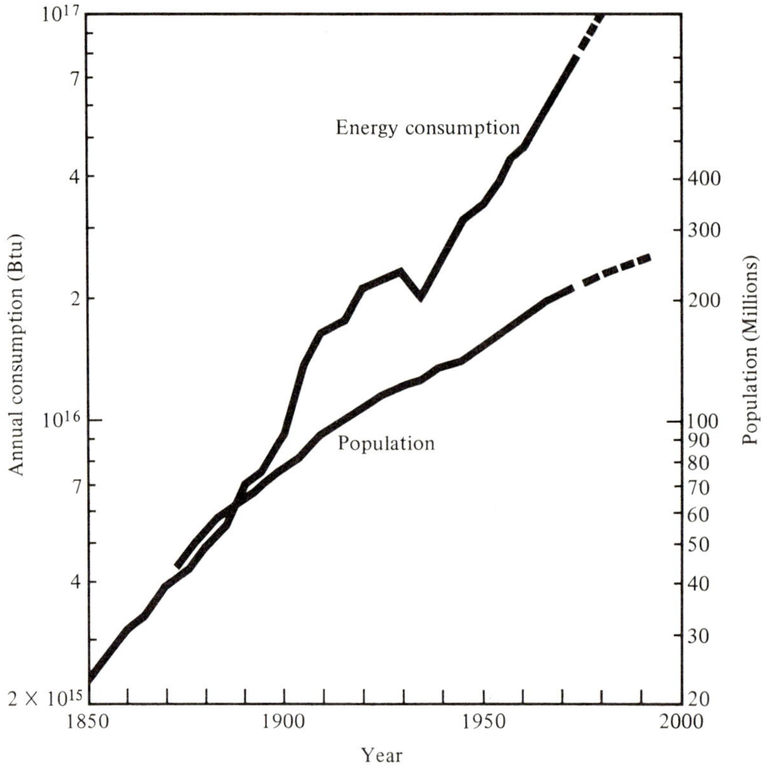

Energy consumption and population growth in the United States from 1850 to 1978.

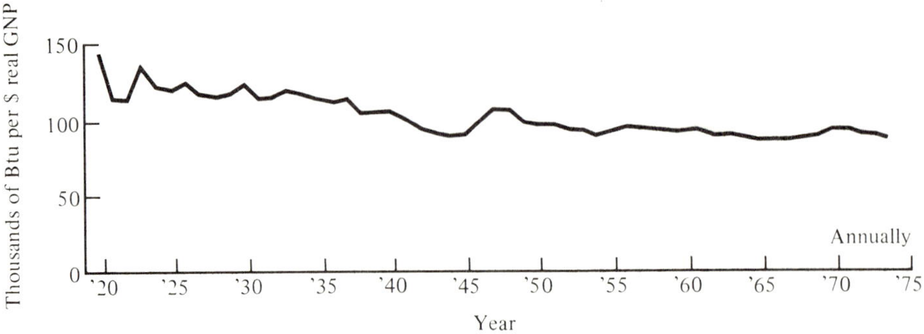

Energy Consumed per Unit of Economic Output (real GNP in 1958 dollars) in the USA. By this measure, U.S. industry is an efficient user of energy.

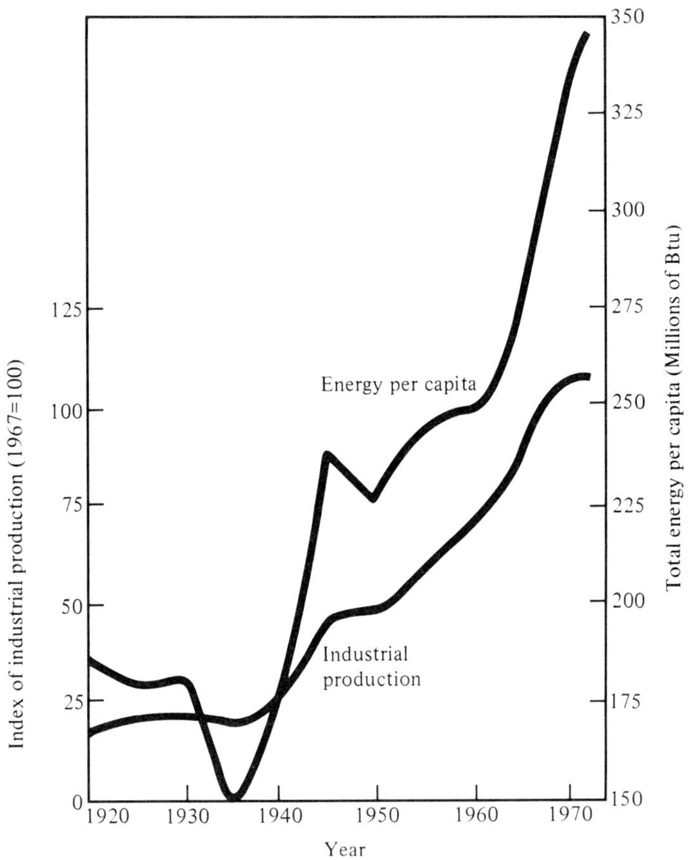

The Growth of Energy per capita use and Industrial Production during the period 1920 to 1972 in the United States.

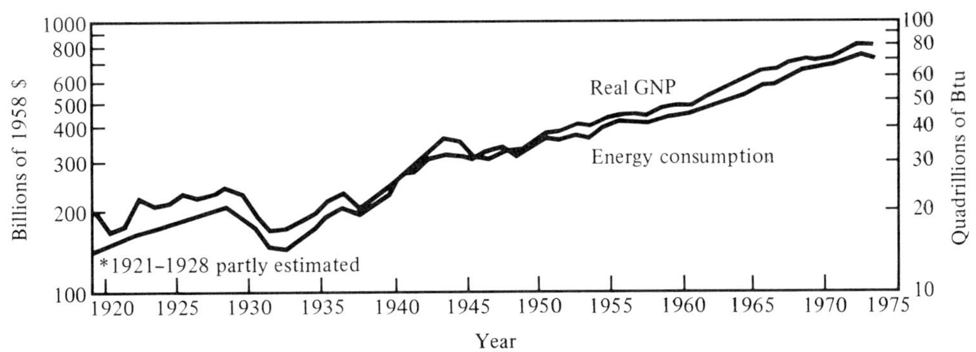

Energy consumption and real GNP as measured in 1958 dollars for the period 1920–1974. Whether GNP pushes up energy consumption or whether energy use causes increased GNP is an interesting and important question.

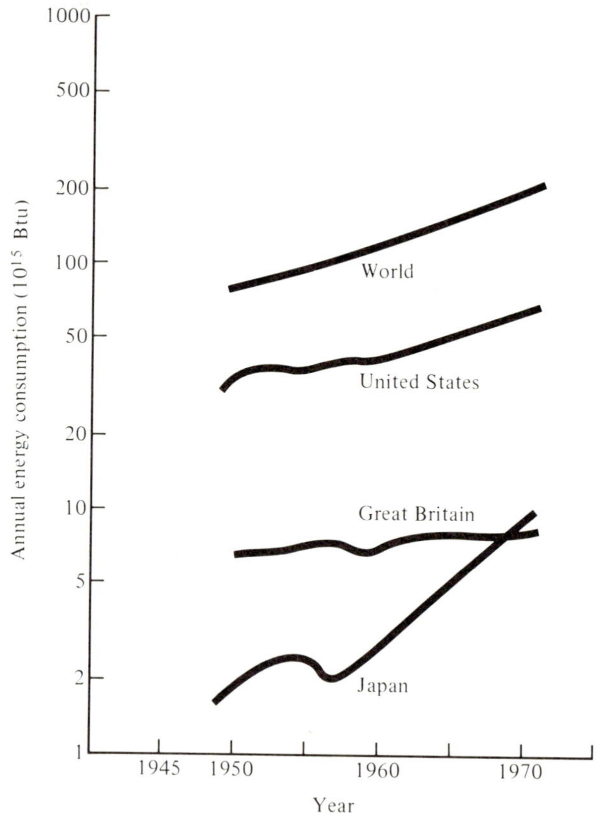

Annual energy consumption for the world, USA, Japan, and Great Britain for the period 1949 to 1971. World energy consumption during the period 1949 to 1971 experienced a rate of growth of 5% per year. Japan has expanded its use of energy during the period 1960 to 1971 at an annual rate of approximately 11%.

CONSUMPTION OF FUEL RESOURCES BY MAJOR CONSUMER GROUP IN THE USA

Percent Distribution

Consumer Sector	*1965*	*1970*	*1973*
Household and Commercial	22.2	20.8	20.3
Industrial	32.3	30.2	28.4
Transportation	23.8	24.5	24.8
Electrical Generation	20.7	24.2	26.2
Other	1.0	0.3	0.3
	100.0	100.0	100.0

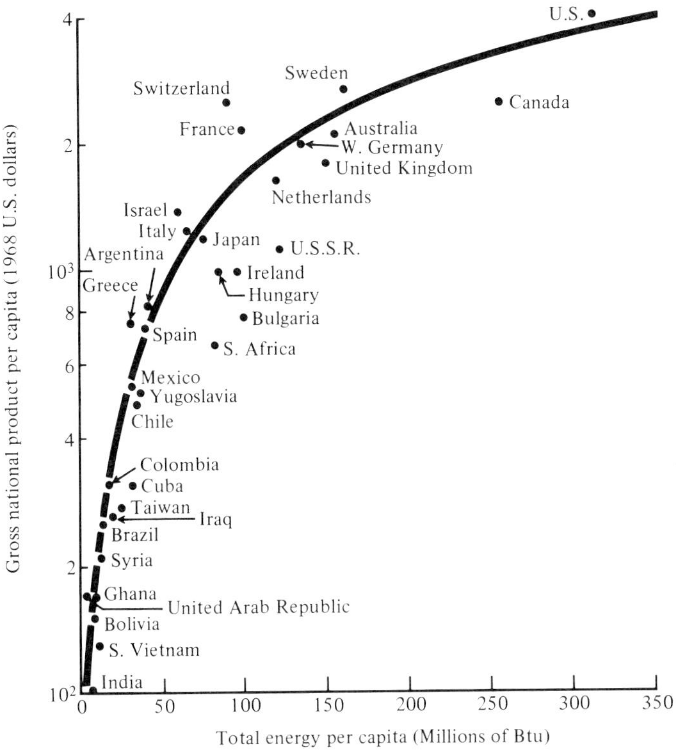

Energy per capita versus the gross national product per capita in 1968 for several nations. There is a trend towards higher consumption as a nation industralizes and increases its gross national product. As a nation enters a stage of affluence, then energy use depends upon supply, price and other factors.

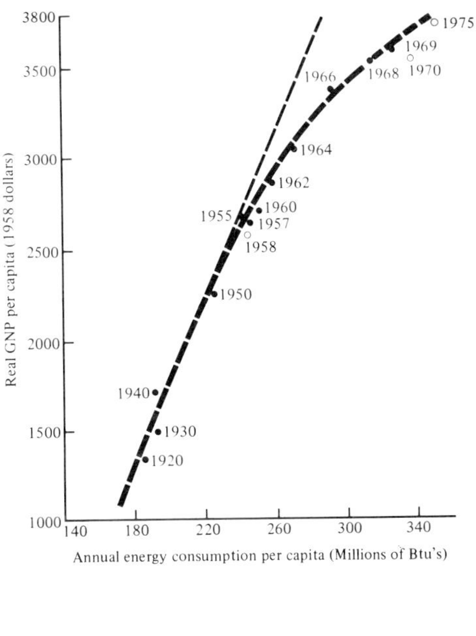

Energy consumption versus real GNP per capita for the period 1920 to 1975, in the U.S.A.

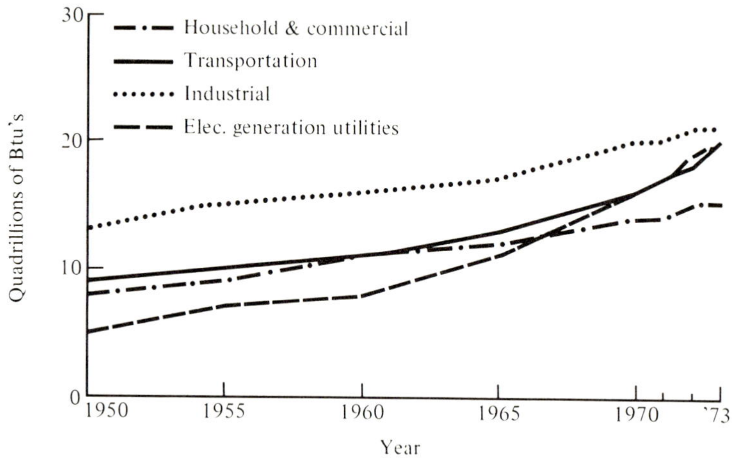

The use of energy in the four sectors of the U.S. market. *Source:* Social Indicators, U.S. General Printing Office, Washington, D.C., 1973.

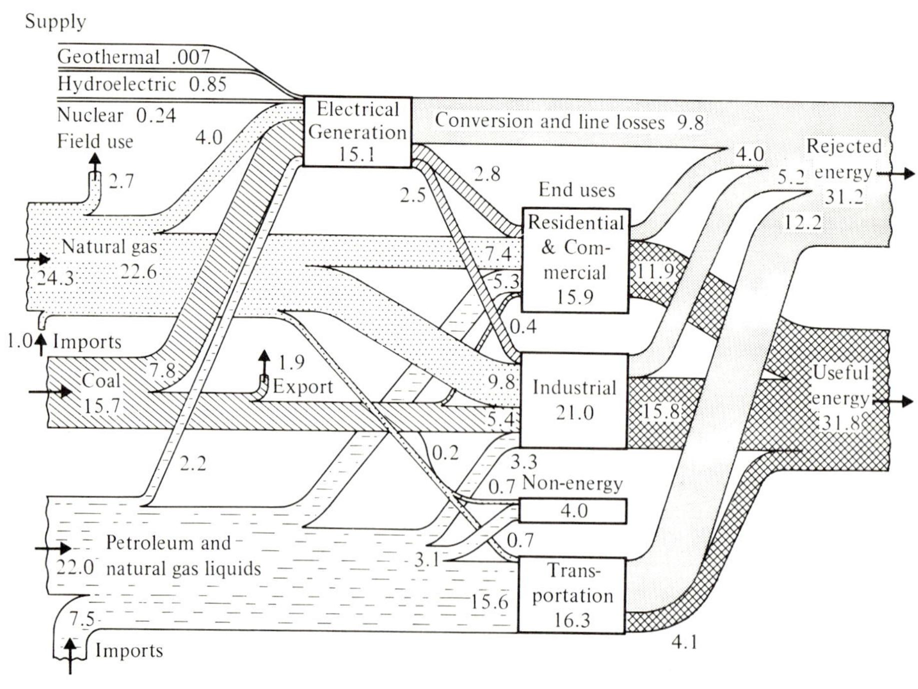

The flow of energy in the U.S. in 1970. All values are given in 10^{15} BTU = 1Q. Total input is 71.6Q. Examining the output side we note that the useful energy, 31.8Q, is only 50.5% of the total output.

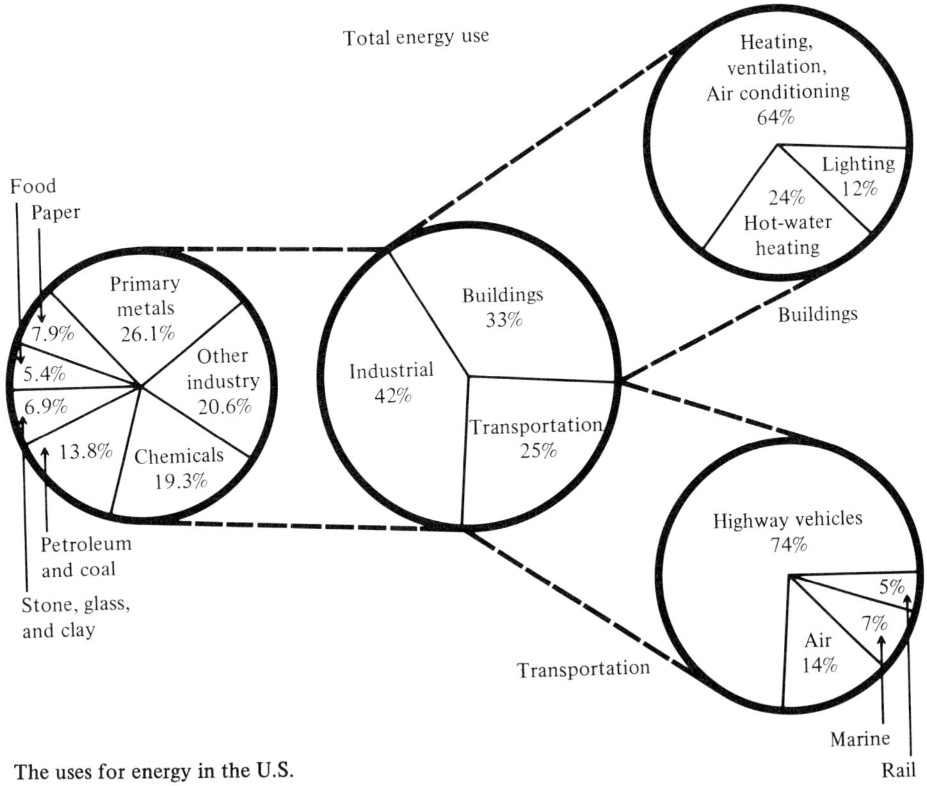

The uses for energy in the U.S.

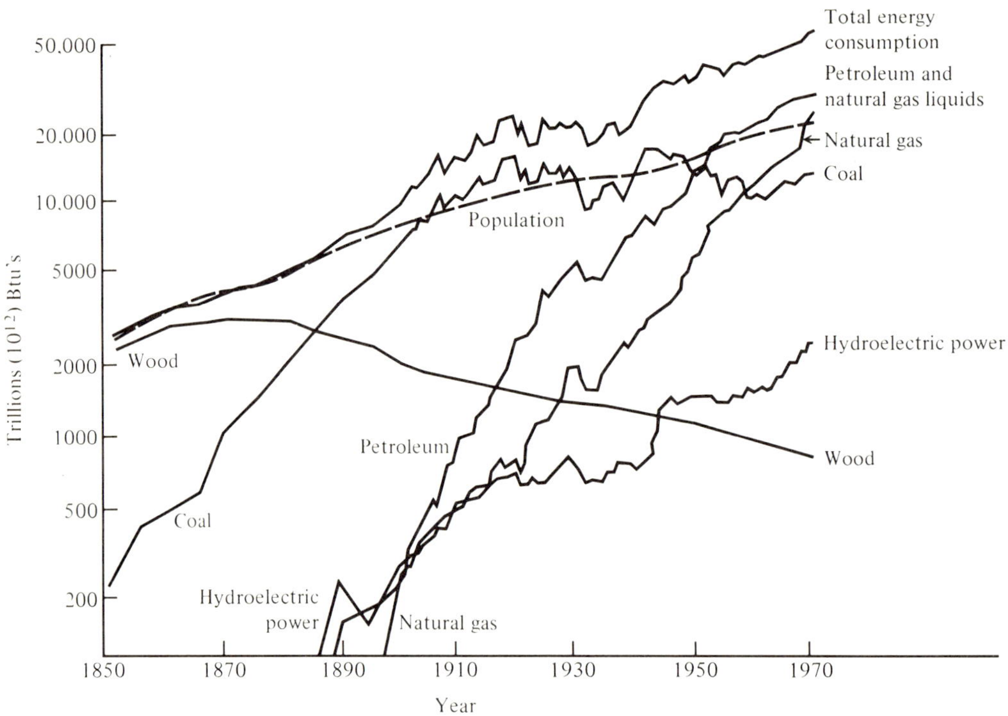

The history of U.S. energy resources since 1850. The dotted line shows the population growth is through that period. Note the vertical scale is logarithmic.

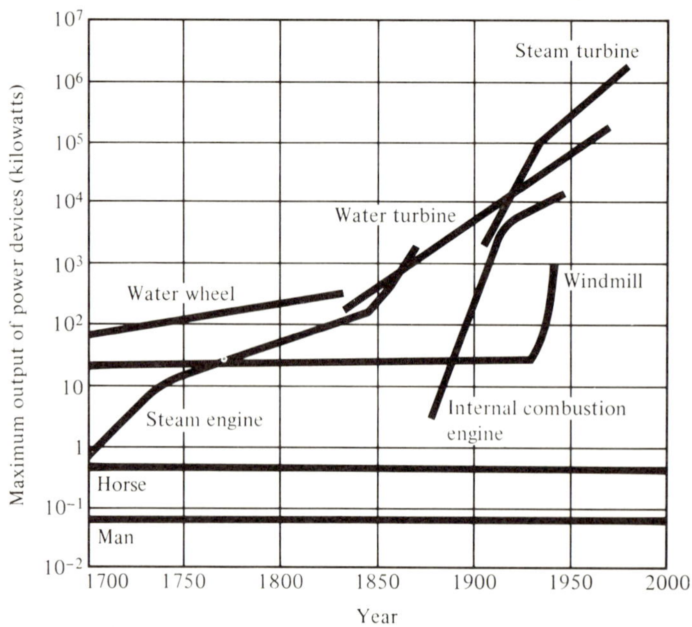

The maximum power output of selected power devices over the period 1700 to 1970.

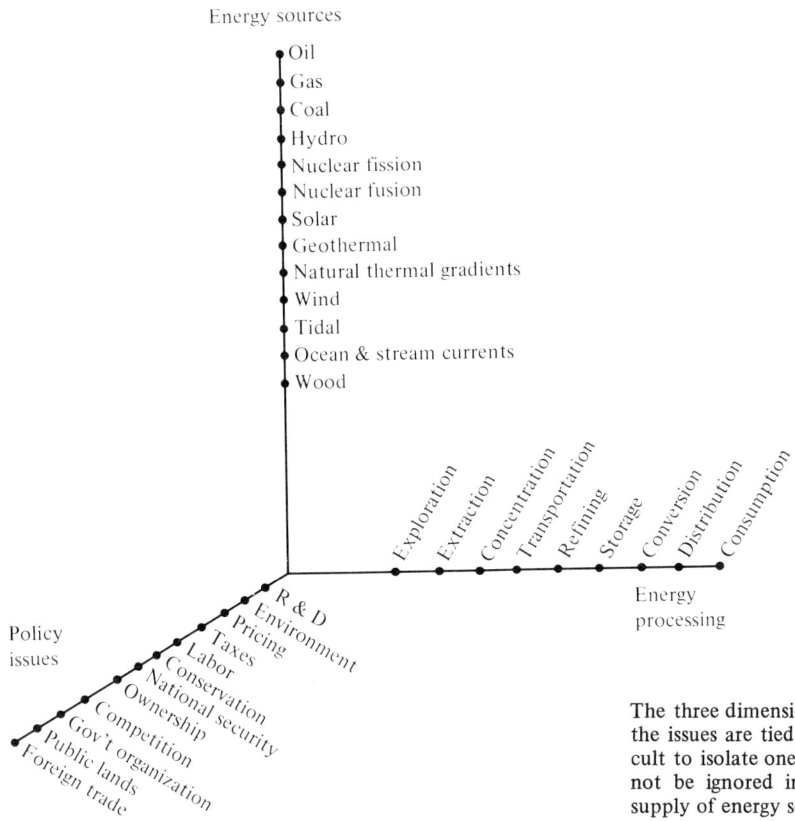

The three dimensions of the energy issue. All of the issues are tied together and it is often difficult to isolate one factor. The policy issues cannot be ignored in favor of the matter of the supply of energy sources.

HISTORICAL SUMMARY OF ENERGY

Year	*Technology*	*Supply*	*Life Style Changes*
1776	James Watt's Steam Engine	Fuel Sources are Wood, Hay, Coal	
1807	Gas Street Lights First Installed in London	Clermont - 1st Commercially Successful Steamship	
1814	1st Practical Steam Locomotive		Gas Lamp Illumination
1831	Henry's Electric Motor	Best Friend Locomotive	

HISTORICAL SUMMARY OF ENERGY (Con't)

	Technology	*Supply*	*Life Style Changes*
1859		Drake Well Hits Oil at 69 Feet	
1863	Open Hearth Furnace		
1867	Refrigerated Rail Car Patented		
1880	Edison Patents Electric Light Bulb		U.S. Railroad Mileage is 87,800
1882		Rockefeller forms Standard Oil Trust	U.S. is Producing 85% of World's Crude Oil
1888	Tesla's First Electric Motor		London Subway Opens
1890	Suez Canel Opens		
1893	Benz and Ford Construct Their First Car	Los Angeles Oil Field Discovered	
1900			Standard Oil Trust Said to Control 80% U.S. Oil
1903	Airplane - Wright's 1st Flight		
1908			Model T Assembly Line
1911	Kettering Develops Electric Self-Starter for Auto	Standard Oil Trust Broken Up	
1914	Ford Develops First Tractor	Crude Oil Worth 88¢/Barrel	
1921			Gasoline Production in U.S. = 472 Million Barrels
1927	Lindberg Flies N.Y. to Paris in 33.5 Hours Nonstop		

	Technology	Supply	Life Style Changes
1929		World's Largest Refinery Built in Arabia	Stock Market Collapse
1931	E.O. Lawrence Invents Cyclotron	Natural Gas from Texas Reaches Chicago Through 24" Pipeline	
1937	1st Jet Engine Built by Frank Whittle		
1942	Enrico Fermi Splits the Atom		
1945	First Atomic Bomb		
1949	USSR Tests its 1st Atomic Bomb	Texas Producing 85% of World's Petrochemicals	British Gas Industry Nationalized
1953	First Hydrogen Bomb		
1957	USSR Launches Sputnik	U.S. Crude Oil Production at 915 Million Barrels/Day	
1962		200 Atomic Reactors in U.S.	
1963	First Oil Drilling Off California Coast		
1965	Northeast U.S. Loses Electricity (Blackout) for Several Hours	Gas Discovered in Alaskan North Slope	
1968			Aswan Dam Completed
1973	Gasoline Rationing or Allocation	Middle East Oil Prices Doubled	
1975		Oil Depletion Allowance Eliminated	China Becomes Oil Exporter

Source: Two Hundred Years of Energy, Chiltons Oil and Gas Energy, January, 1976.

3

PROJECTIONS OF ENERGY CONSUMPTION

The continued growth of the use of fossil fuels can occur only for a limited number of years before these resources are exhausted. The implications of compound growth often escape our attention until it is too late. For example, in the U.S. we have used about 100 billion barrels of our domestic oil resource, and there may be 100 billion barrels remaining as recoverable resources. Since we have used oil for about 100 years, it may be tempting to assume that oil should last for another 100 years. But the fact is that the consumption of oil has been increasing at a rate of 5% per year, and the remaining oil could be depleted in another 14 years.

DOUBLING PERIOD FOR SPECIFIED GROWTH RATES

Rate, r	0.01	0.03	0.05	0.06	0.07	0.09	0.10	0.12
Period to double, years	69.6	23.4	14.2	11.9	10.2	8.0	7.3	6.1

Compound growth at a rate r (a decimal) will result in a doubling of the quantity in a specified number of years. As an approximation, the doubling period is $0.72/r$. With exponential consumption of a resource, the amount of a product consumed during one doubling period is equal to the total used over all preceding time.

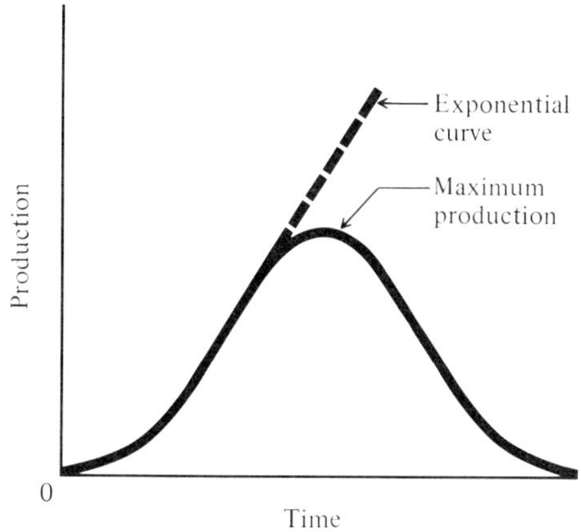

The production of a finite fossil fuel. The production of an exhaustable fossil fuel may follow a curve as shown. The production increases following an exponential curve passes through a maximum and declines eventually to a very small quantity as the fuel approaches extinction.

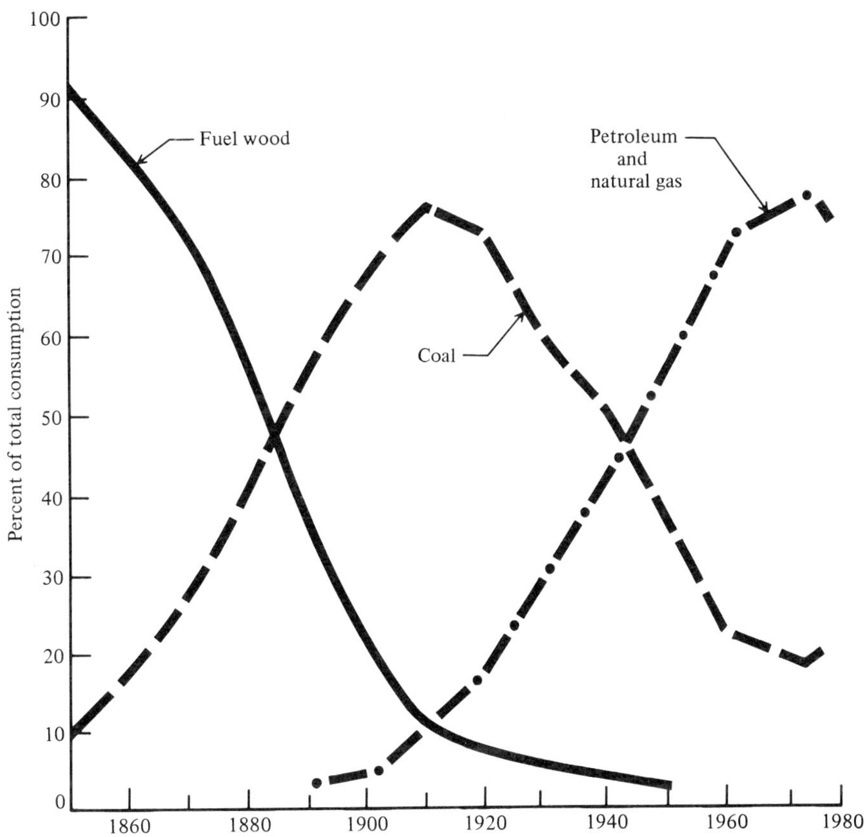

The utilization of wood and coal as a fuel exhibit the production pattern as illustrated in Fig. 3-1. It is expected that petroleum and natural gas production will decline in the U.S. over the next two decades.

The consumption of a limited fuel over the lifetime of the fuel. The finite quantity of the fuel is Q_f and the time is t_f when the fuel is exhausted. If we define Q_R as the proven reserves and Q_p as the total accumulated quantity of fuel removed from the ground, the cumulative discoveries Q_d, represents the quantities removed from the ground plus the recoverable quantities remaining in the ground. Therefore,

$$Q_d = Q_p + Q_R$$

Since Q_d is eventually limited to Q_f we may represent the changes in these quantities as shown.

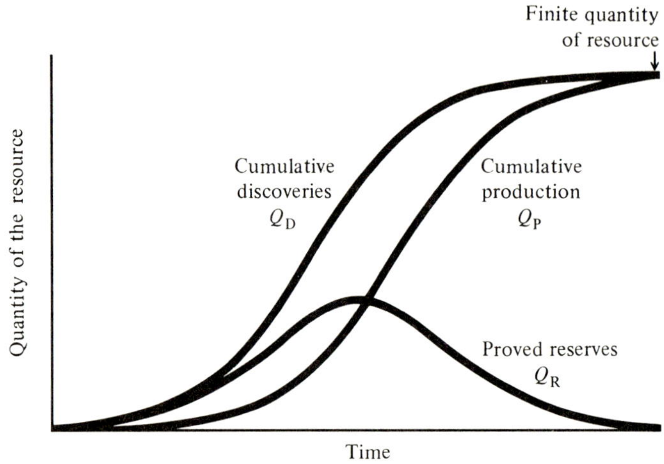

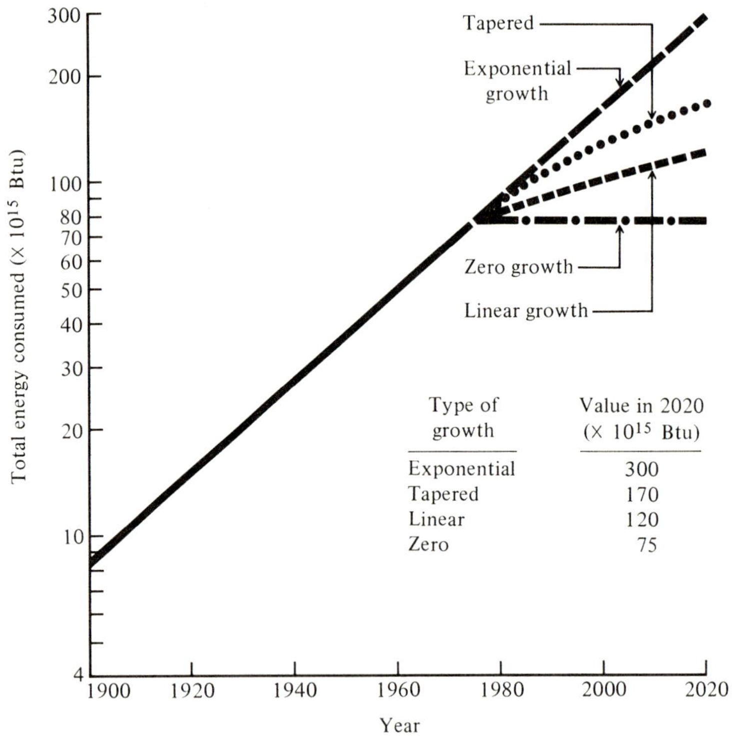

The historical growth of energy consumed in the U.S. and several projections for the period 1975 to 2020. The curve of the historical exponential growth is extrapolated to the year 2020 yielding a value of 300×10^{15} Btu for that year. A curve for zero growth is also shown. The curve for linear growth following an increase of 1×10^{15} Btu's per year is also shown. We can also calculate a curve for a tapered growth resulting from a decleration of the use of energy in the U.S. A curve is shown which follows a linear deceleration of the growth rate to a zero growth rate by 2025. We can expect that many of the curves related to energy consumption in the world will experience a tapered growth as prices of fossil fuels increase and supply falls behind demand.

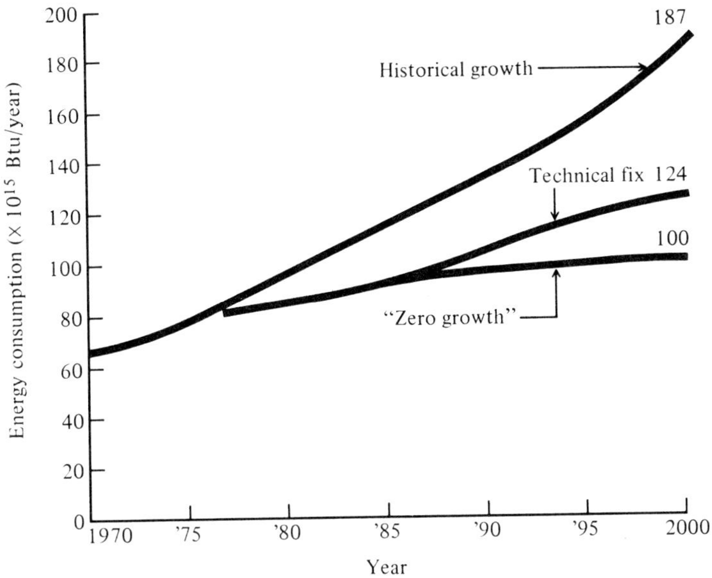

Three forecasts of energy consumption in the U.S. during the period 1975 to 2000. These forecasts are a result of the Ford Foundation Energy Policy Project. The first forecast is based on the assumption that energy use in the U.S. will continue to grow until the end of the century at about 3.4% annually, the average rate of growth from 1950 to 1970. The second forecast is based on the assumption that a conscious national effort to use energy more efficiently through the introduction of energy conserving technology is called the *technical fix* scenario. The Ford Foundation project states that if these technical approaches were consistently applied, an energy growth of 1.9% per year would be adequate to satisfy our national needs. The Ford project proposes zero energy growth scenario. While called zero growth, it actually has decreasing growth rates eventually reaching zero after 2000. The scenario assumes a growth rate of 1.76% from 1975 to 1985 and 0.47% over the period 1985 to 2000. *Source: A Time to Choose,* Report of the Energy Policy Project of the Ford Foundation, Ballinger Pub. Co., Cambridge, Mass., 1974.

THREE FORECASTS OF THE FORD ENERGY PROJECT

	Energy Consumption (X 10^{15} *Btu*)		
	1975	*1985*	*2000*
Historical Growth	78	116	187
Technical Fix	78	91	124
"Zero Growth"	78	93	100
Tapered Growth of Figure Above	78	100	132

PROJECTIONS FOR THE YEAR 2000 FOR THE TOTAL ENERGY CONSUMPTION IN THE U.S.

	Consumption *(X 10^{15} Btu)*
1. National Energy Plan (1977)	125
2. Electric Power Research Institute (1979)	130
3. Earl T. Hayes (1979)	95
4. von Hippel/Lovins (1977)	89–95
5. U.S. Dept. of Energy (1979)	125

Sources: 1. Dept. of Energy
2. EPRI Journal, May 1979
3. E. T. Hayes, Science, Jan. 19, 1979, pg. 234.
4. Power Engineering, May 1979, pg. 25.
5. Dept. of Energy, Energy Information Administration, April 1979.

INCREASE IN ENERGY USE FOR 1951 TO 1969

	Energy Increase *(X 10^{12} Kilowatthours)*	*Percent Increase*
World total	37.2	100%
Third World	4.1	11%
USSR and Eastern Europe	13.0	35%
USA and Western Europe	20.1	54%

The growth rate of energy consumption by the world followed an exponential rate of 4.5% over the period 1950–1970.

A PROJECTION OF WORLD ENERGY CONSUMPTION FOR THE PERIOD 1970 TO 2000

Year	*World Energy Consumption* *(X 10^{15} Btu)*
1970	216.9
1980	300
1985	351
1990	417
2000	522

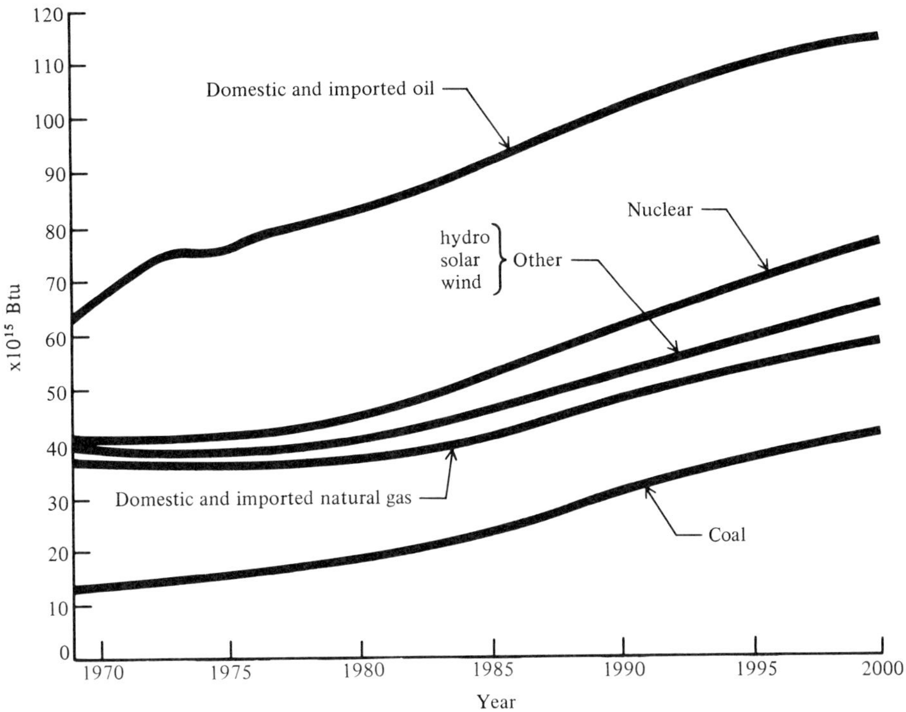

A graphic portrayal of the projection for energy consumption for the U.S. for the period 1972 to 2000 as detailed.

ONE PROJECTION FOR ENERGY CONSUMPTION IN THE U.S. FOR THE PERIOD 1972 TO 2000 IN QUADS

Domestic Supply	*1972*	*1977*	*1985*	*1990*	*2000*
Coal	14.5	15.9	22.7	31.2	41.7
Gas	22.2	19.6	18.2	17.4	17.1
Oil	22.6	19.8	21.8	22.0	23.0
Nuclear Fission	0.6	2.7	6.6	9.1	11.6
Other (Hydro, Wind, Solar)	2.9	2.4	3.2	3.8	4.6
Domestic Production	62.8	60.4	72.5	83.5	97.8
Gas Imports	1.0	1.0	1.9	2.0	2.0
Oil Imports	9.8	18.2	17.4	16.0	15.0
TOTAL SUPPLY	73.7	79.8	91.8	101.5	115.0

This projection is based on an emphasis on increased coal use during 1980–2000 and a reduction of dependency on imported oil. It may be possible to increase the supply of energy from solar, wind, and hydroelectric sources.

4
NATURAL AND SYNTHETIC GAS

Natural occurring gas and gas manufactured from coal or oil has been used for lighting, cooling, and heating fuel for over a century. In 1880 the world used approximately 200 billion cubic feet of fuel gas. Almost one century later, the world is using three hundred times that amount of gaseous fuel.

The use of natural gas or gaseous fuel generated from coal came into use in the U.S. after 1890. In the U.S. in 1900 approximately 200 billion cubic feet of gas was used as an energy fuel. Today, about 22 trillion cubic feet of gas is used as a fuel in the U.S., an increase of over one hundred times the amount used 75 years ago.

Natural gas was discovered in the U.S. along with petroleum and became available as petroleum wells were developed. Since it was not as easily stored or transported as liquid petroleum, natural gas, which is CH_4 (methane), was not generally used at first. However, as it was recognized as a valuable fossil-fuel, methods of storage and transportation were developed.

Natural gas can be found with oil in associated gas fields or independently from oil in unassociated gas fields. In the past some gas has been generated from oil and some from coal. Gas is the cleanest, in terms of pollutants, and most flexible natural fuel, but it is more difficult to store and transport than liquid or solid fuels, especially from one continent to another.

Synthetic Fuel (Synfuel): A commercially-desirable fossil fuel, normally oil or gas, obtained as a result of a conversion of a fossil fuel available in a less useful form to the commercially desirable form. For example, the conversion of coal to synthetic gas or the conversion of oil shale to petroleum.

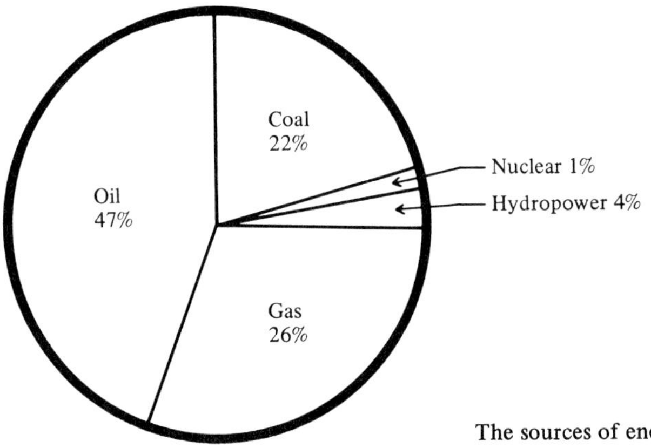

The sources of energy in the U.S. in 1978.

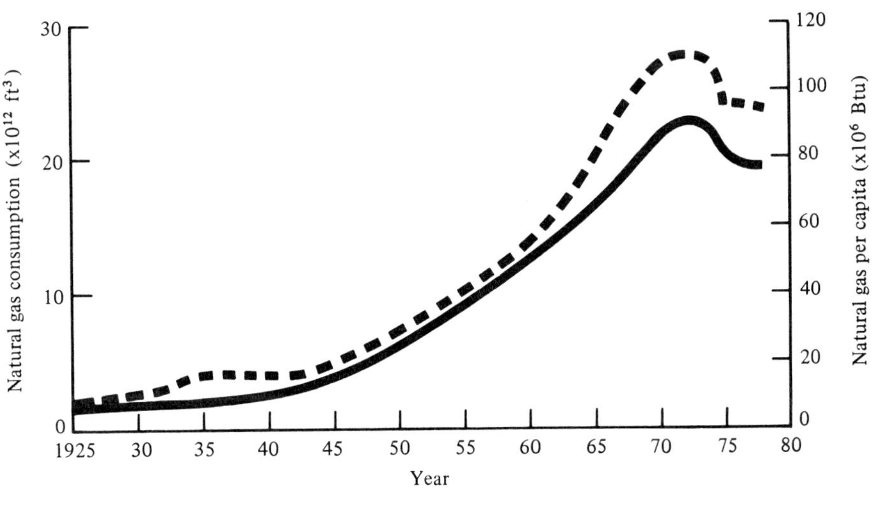

The annual natural gas consumption (solid line) in trillions of cubic feet and per capita natural gas consumption in millions of BTU (dashed line) for the USA for the period 1925 to 1978.

USE OF NATURAL GAS IN THE U.S. FOR THE PERIOD 1920 TO 1978

Year	*Annual Gas Consumption (Billions of ft^3)*	*Gas Consumption Per Capita (Millions of Btu)*	*Average Wellhead Price (Cents/1000 ft^3)*	*Index of Real Price 1967 = 100 (Constant Dollars)*
1920	812	8		
1925	1,209	10		
1930	1,979	16		
1935	1,969	16		
1940	2,734	21		
1945	4,042	30		
1950	6,282	41		
1955	9,405	56	10.4	75
1960	12,771	71	14.0	92
1965	16,040	83	15.6	101
1970	21,921	108	17.1	97
1972	22,532	110	18.6	98.6
1973	22,648	109	21.0	103
1974	21,712	108	30.0	110
1975	20,000	96	43.0	130
1977	19,521	93	63.0	170
1978	19,372	92	77.0	195

PER CAPITA CONSUMPTION OF NATURAL GAS AND TOTAL ENERGY FOR 1920 TO 1980 IN MILLIONS OF BTU

Year	*Natural Gas*	*Total Energy*	*Percent Natural Gas of the Total*
1920	8	186	4
1930	16	181	9
1940	21	181	12
1950	41	226	18
1960	71	249	28
1970	110	338	33
1978	92	315	29

CONVERSION FACTORS FOR NATURAL GAS

1 cubic foot of natural gas = 1031 Btu's = 1.088×10^6 joules
1 cubic foot of natural gas = 0.0283 cubic meters = 28.3 liters

THE DEVELOPMENT OF A U.S. INTER-STATE TRANSMISSION PIPELINE SYSTEM

Year	*Gas Consumption (10^{12} Cubic Ft.)*	*High Pressure Pipelines (Thousands of Miles)*	*Percent Gas Sold Outside Producing States*
1945	4.0	77	48
1950	6.3	109	52
1955	9.4	142	55
1960	12.7	181	60
1965	16.0	210	64
1970	21.9	253	67
1974	21.7	265	68

The transportation gas via pipeline is a relatively inexpensive means of transmitting energy. Natural gas can be transmitted via pipeline for 3 cents/10^6 Btu over a one-hundred mile distance. A typical pipeline is 48 inches in diameter.

CONSUMPTION OF NATURAL GAS IN THE FOUR SECTORS FOR THE PERIOD 1947 TO 1978 IN BILLION CUBIC FEET

Year	*Household and Commercial*	*Industrial*	*Transportation*	*Electricity Generation*
1947	1,087	2,906	Negligible	373
1950	1,586	3,602	126	629
1955	2,753	4,769	245	1,153
1960	4,123	6,074	347	1,724
1965	5,346	7,433	501	2,318
1970	6,894	9,856	722	3,894
1971	7,346	10,438	825	4,125
1975	7,190	8,370	580	3,160
1978	6,300	8,200	520	3,220

Source: U.S. Dept. of Energy

WORLD CONSUMPTION OF NATURAL GAS BY REGIONS AND NATIONS IN 1974

Nation	*Natural Gas (10^{15} Btu)*	*Total Energy (10^{15} Btu)*	*% Natural Gas of Total Energy*
USA	21.7	73.9	29
Canada	2.4	7.0	34
USSR	9.0	44.0	20
Japan	0.13	12.4	1
Western Europe	3.7	45.7	8
Eastern Europe	1.3	13.7	10
World Total	52.0	250.0	20.8

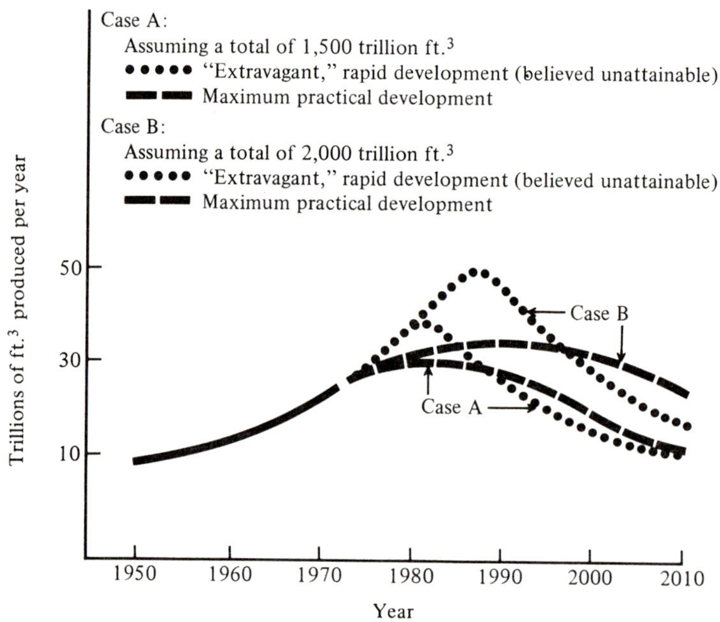

A portrayal of 4 possible consumption patterns for the use of the natural gas resources of the U.S. The dotted lines show a rapid, but probably unattainable use of the gas. The dashed lines show a realistic possible use for each of the two ultimate potentials for gas. Case A assumes a total resource of 1,500 trillion ft^3 while case B assumes a total resource of 2,000 trillion ft^3.

KNOWN RESERVES OF NATURAL GAS IN THE U.S.

Year	*Known Reserves (X 10^{12} Cubic Feet)*	*Reserve to Production Ratio*
1960	260	20.2
1965	285	17.5
1967	289	15.5
1970	284	12.5
1972	238	11.5
1973	218	10.5
1974	237	11.1
1975	228	11.4
1978	200	10.1

ESTIMATES OF WORLD RESOURCES OF NATURAL GAS IN TRILLIONS OF CUBIC FEET

Nation or Region	*Total Resource Q_T*
U.S.	1,500
Canada	600
Middle East	3,500
Far East	1,300
South America	1,400
Western Europe	300
Africa	1,600
USSR and Eastern Europe	2,800
World Total	13,000

THE AVERAGE PRICE OF NATURAL GAS SOLD TO U.S. INDUSTRY

Year	*1965*	*1970*	*1975*	*1978*
Pricc (Dollars per Million Btu)	.30	.31	.90	2.0

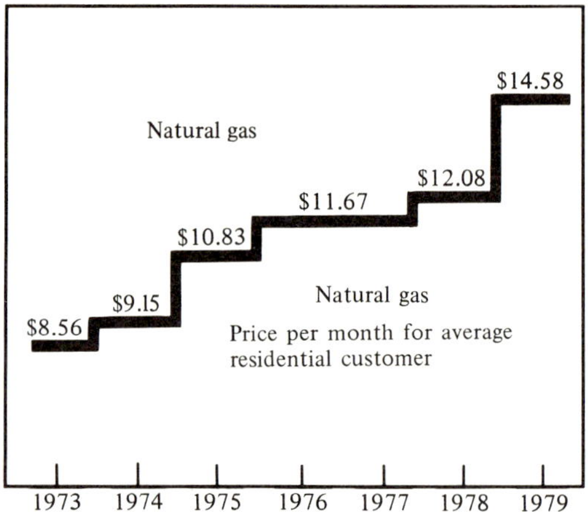

The price of natural gas during the period 1973 to 1979 delivered to an average residential customer in Southern California.

5
COAL

After decades of declining production and increasing disfavor, coal, the most abundant energy resource in the United States, has been making a strong comeback. It is one of the ironies and dilemmas of our environmentally-aware age that in the future we will use more, not less, of this relatively heavy polluting fuel.

Coal, the fossil fuel extensively used by man, was initially used by the Chinese at the time of Marco Polo or earlier. The use of coal as a major source of energy began in England in the twelfth century, and it was deduced that coal could be dug from strata of rock along the cliffs in England and then from holes sunk to the strata. By the late 13th century, coal smoke was already a source of pollution in London. Coal was used as a domestic fuel, a fuel for lime burning, and by blacksmiths and others for metallurgical processes. By 1750 the annual production in England reached 7 million tons.

Coal was used during the industrial revolution for metallurgical processes, glass-making fuel for railroads, and in general for the steam engine. By 1860, world production of coal reached 150 million tons. From the period 1860 to 1910, annual world production of coal grew from 150 to 1,100 million tons, at an annual growth rate of 4.4 percent.

Coal is actually a family name for a variety of fuels. The beginnings of coal were plants which were accumulated in a bog and become a soggy mass of plant debris we call peat. When peat was compressed over 300 million years ago, it became lignite. Successive invasions of the sea and piling on of layer upon layer of material resulted in the deep burial of the lignite. Deep burial often results in a rise in temperature and an expelling of the moisture, and thus lignite became bituminous coal. In some areas the layers of coal were subjected to large compressive forces thus resulting in "hard coal" or anthracite. The main constituent of coal is carbon and hydrogen, with added small granites of sulfur, oxygen, and nitrogen.

U.S. ANNUAL COAL PRODUCTION

Year	*Production (Millions of Tons)*
1850	10
1890	160
1910	500
1930	500
1935	350
1940	510
1945	600
1947	630
1950	516
1960	440
1965	513
1970	590
1975	635
1978	660
1985 (Projected)	800

WORLD COAL PRODUCTION

Year	*Production (Millions of Metric Tons)*
1860	150
1880	320
1900	780
1920	1,200
1940	1,700
1960	2,100
1970	2,500
1977	3,400

AVERAGE PRICE OF A TON OF COAL SOLD TO ELECTRIC UTILITIES

Year	*1973*	*1974*	*1975*	*1976*	*1977*	*1978*
Price (Dollars per Ton)	10	13	17	20	23	25

CONVERSION FACTORS FOR COAL

One ton of coal (2000 lbs.) yields 25 million Btu

One ton of coal = 25,000 cubic feet of natural gas
= 189 gallons of gasoline

One ton (2000 lbs.) = 907 kilograms
= .907 metric ton

INTERNATIONAL COAL PRODUCTION IN 1977

Country	*Production (Millions of Metric Tons)*	*Percent of World Production*
USSR	730	21.0
USA	650	18.4
China (PRC)	510	14.6
East Germany	250	7.4
Poland	230	6.6
West Germany	200	6.1
United Kingdom	110	3.6
Czechoslovakia	100	3.5
TOTAL WORLD	3,400	100.0

USES OF COAL FOR VARIOUS PURPOSES IN THE U.S. IN 1978

	Percentage of Total Use
Electric Utility	65.0
Industrial Steam	15.9
Coking Coal for Steel	10.1
Railroad Fuel	0.2
Residential and Commercial Heating	1.5
Export	7.3
TOTAL	100.0

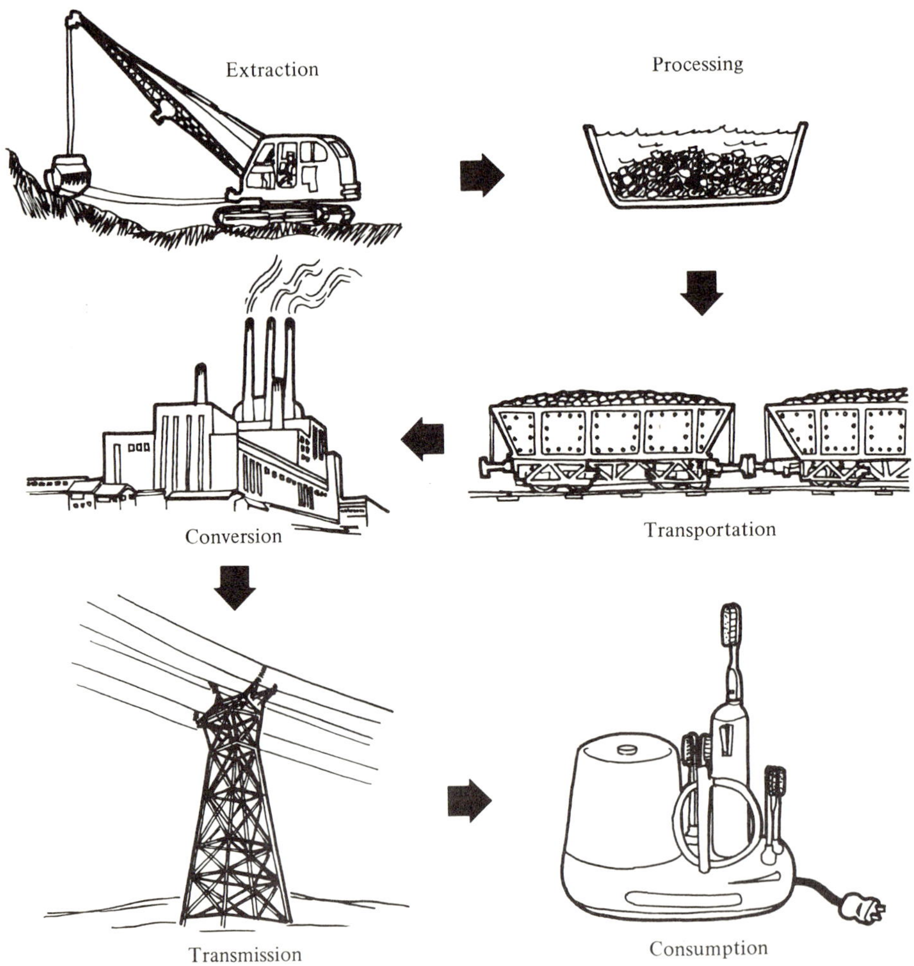

The process of coal utilization for electric power generation.

WORLD COAL RESERVES (IN UNITS OF 10^{18} Btu)

	Proved	*Estimated Total*
U.S.A.	40.0	83.4
Canada	16.0	37.0
South America	0.4	0.8
West Europe	10.0	21.6
East Europe and U.S.S.R.	120.0	250.0
Africa	3.0	6.2
TOTAL	189.4	399.0

Note: World consumption of total energy is about 175×10^{15} Btu per year.

U.S. ESTIMATED COAL RESOURCES

Depth of Overburden (Feet)	*Type*	*Resources (Billions of Tons)*	*Energy Reserve ($\times 10^{18}$ Btu)*
100	Strip Coal	140	3.6
100 to 3,000	Bituminous	959	
	Lignite	448	37.0
	Anthracite	13	
3,000 to 6,000	All Types	337	8.8
6,000 to 9,000	All Types	1,313	34.1
		3,210	83.5

Note: Current mining methods are not economical below depths of 1,000 feet.

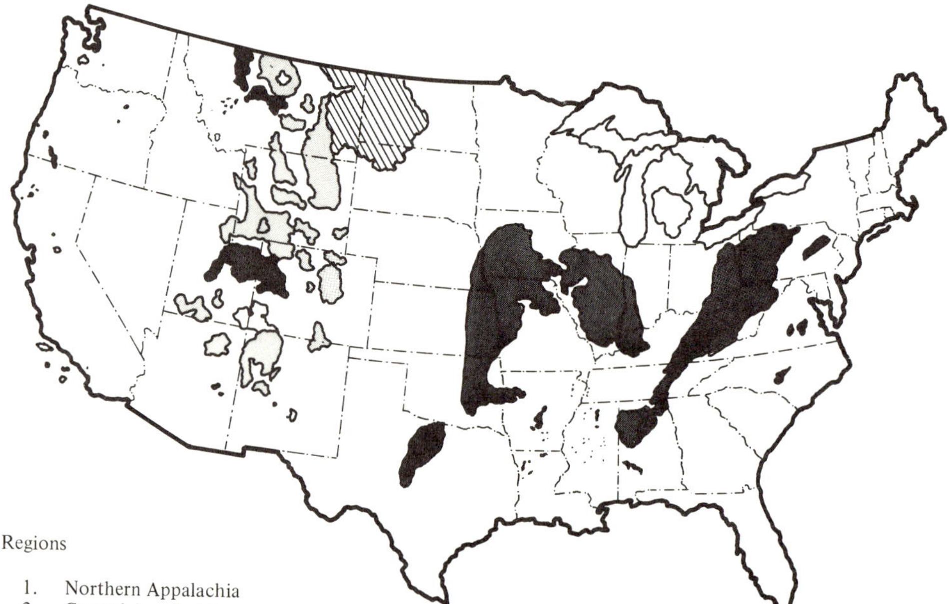

Regions

1. Northern Appalachia
2. Central Appalachia
3. Southern Appalachia
4. Midwest
5. Central West
6. Gulf
7. Eastern Northern Great Plains
8. Western Northern Great Plains
9. Rockies
10. Southwest
11. Northwest
12. Alaska (not shown)

Legend

Bituminous coal

Sub-bituminous coal

Lignite

Anthracite

Coal Supply Regions *Source: 1976 National Energy Outlook,* Federal Energy Administration

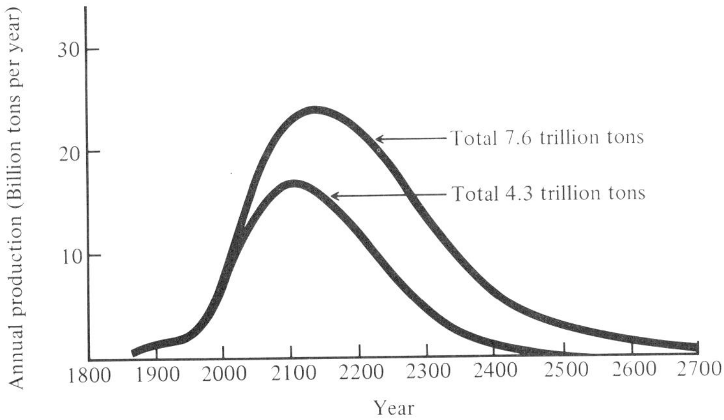

The rate of production of coal in the world for two estimates of total resources, Q_{total}.

SELECTED COAL GASIFICATION PROCESSES

Name of Process	*Heat Input*	*Reactor Type*	*Status*
Lurgi	Oxygen	Downward-Moving Bed	Near Demonstrated
Hygas	Oxygen	Fluidized Bed	Pilot Plant
Bi-Gas	Oxygen	Entrained Flow	Will Build Pilot Plant
CO_2 Acceptor	Air	Fluidized Bed and Dolomite	Pilot Plant
Cogas	Air		Bench-Scale Data
Synthane	Oxygen	Entrained/Fluidized	Will Build Pilot Plant

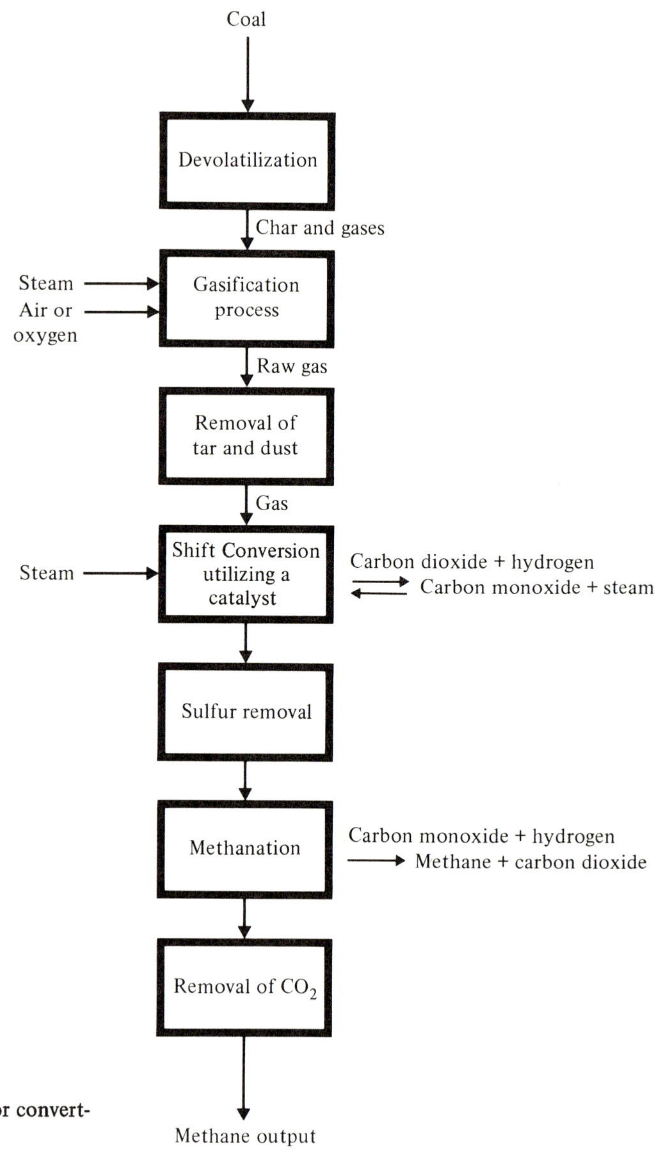

A Schematic of the Coal Gasification Process for converting coal to methane.

The HYGAS pilot plant is operated by the Institute of Gas Technology. The pilot plant tests the HYGAS process for making synthetic gas from coal. The plant is designed to process 75 tons of coal a day to make 1.5 million cubic feet of synthetic gas. A commercial sized plant would use 17,000 tons of coal a day to produce 250×10^6 cubic feet of gas. A commercial plant might produce gas at $4/million Btu. Courtesy of the Institute of Gas Technology.

6
PETROLEUM

Petroleum has assumed an important and growing role as an energy source in the world. The superior qualities and easy convenience of transportation have made petroleum the preferred fuel since 1920. The result is a U.S., European, and Japanese economy heavily dependent upon the availability of oil. The world economy is rapidly approaching the same dependence that industrialized nations now experience.

Nevertheless, we understand that oil, like natural gas and coal is finite in supply. The increased use of petroleum is causing prices to rise around the world. The U.S. has become increasingly reliant on oil as a primary fuel as well as a supply for the production of chemicals and other materials.

ANNUAL CONSUMPTION AND PRODUCTION OF PETROLEUM IN THE U.S.

Year	*Consumption (Millions of Barrels)*	*Domestic Production (Millions of Barrels)*	*Imports as Percent of Consumption*
1920	434	443	2%
1930	862	898	4%
1940	1,285	1,353	5%
1950	2,375	1,974	17%
1960	3,611	2,575	29%
1970	5,365	3,517	34%
1974	5,900	3,500	40%
1978	6,800	3,480	47%
1980 (Projected)	6,800	3,500	45%

U.S. PETROLEUM MILESTONES

Pioneer Period

1859 - First U.S. oil well drilled by Edwin Drake in Pennsylvania.
1905 - Peak of Rockefeller's Standard Oil Trust Petroleum supplies 10% of U.S. energy.
1925 - U.S. produces 71% of the world's oil.
1930 - An oil glut and oil sells for 10¢ per barrel.
1945 - U.S. uses 4.5 million barrels per day.

Growth Period

1953 - U.S. oil production accounts for one-half of world oil production and U.S. consumes almost all of it.
1955 - U.S. has 20% of the estimated oil reserves.

Maturity

1961 - U.S. proven reserves peak and begin to decline.
1970 - U.S. production peaks and begins a decline.
1973 - U.S. consumes 30% of the world's oil supply.
1973 - U.S. imports 47% of its oil consumption.

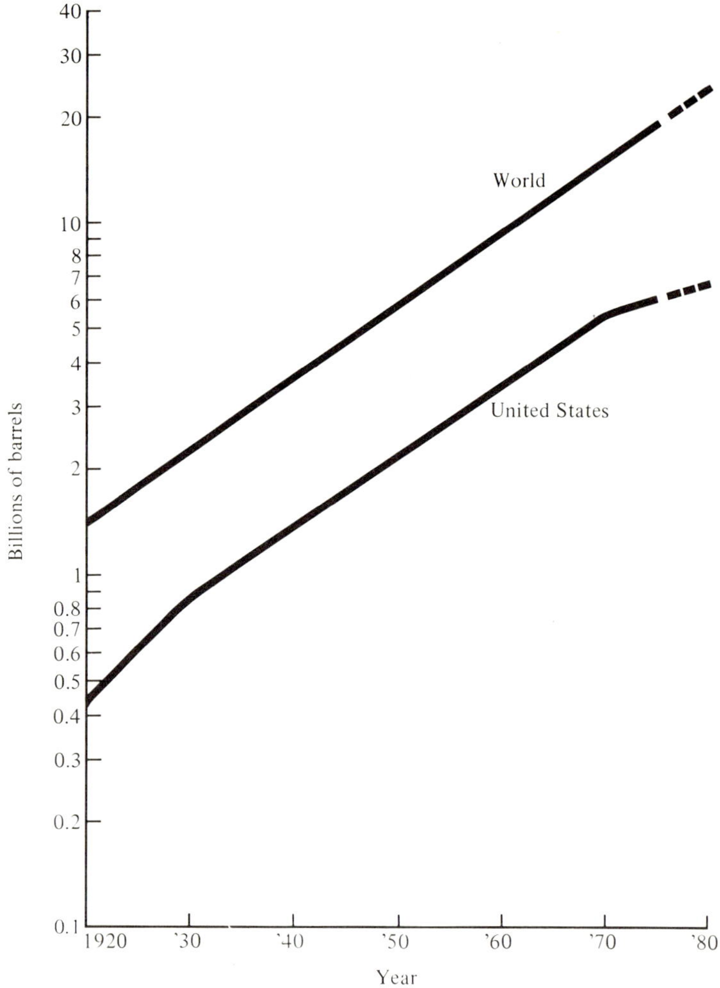

Annual Consumption of Petroleum

PETROLEUM USES IN DECEMBER 1978 IN THE U.S.

Type of Use	*Consumption* (10^{15} *Btu*)	*Percent of Total*
Residential and Commercial	0.614	18.9
Industrial	0.597	18.4
Transportation	1.685	52.0
Electric Utilities	0.348	10.7
TOTAL	3.244	100%

Autos and trucks use 84% of oil consumed in transportation. Highway driving accounts for 45.6% of all oil usage.

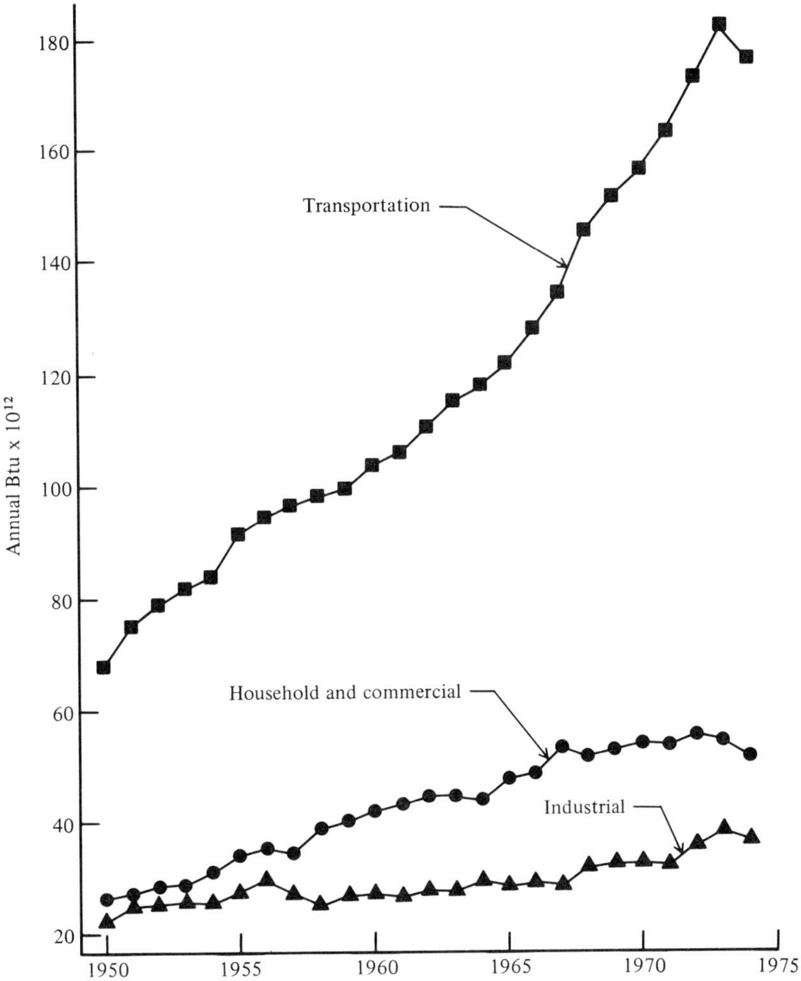

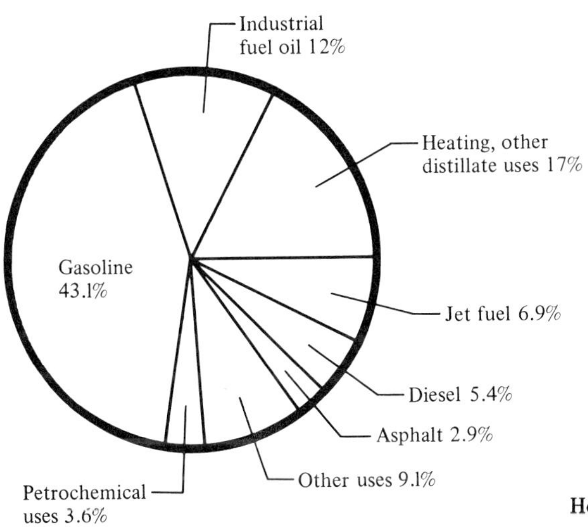

How U.S. Uses Oil

DEMAND FOR PETROLEUM IN THE INDUSTRIAL SECTOR

Year	*Fuel Use (Million Barrels)*	*Non-Fuel Use (Million Barrels)*	*Total (Million Barrels)*	*Industrial Use as Percent of Total Use of Oil*
1950	362	79	441	19%
1960	444	200	644	18%
1970	547	415	961	17%

An oil refinery turns crude oil into gasoline, diesel oil, lubricating oil, kerosene, and jet fuel. Gasoline is the chief product of a refinery (45% average) and fuel oil ranks second in importance. Fuel oil is used to heat homes and buildings and to supply power for factories and railroads. Jet fuel contains a mixture of gasoline, kerosene and oils with low freezing points. Asphalt for roads is another product of crude oil. Non-fuel use of petroleum has grown from 18% in 1950 to 43% in 1970. Non-fuel use of petroleum in the chemical industry is an important use of oil. Petrochemicals are chemicals made from petroleum and examples are plastics, nylon, fertilizers, and medical drugs.

WORLD PRODUCTION OF PETROLEUM IN 1975

Nation or Region	*Annual Production (Billions of Barrels)*
United States	3,720
Canada	694
Latin America	1,752
SUBTOTAL	6,166
W. Europe	183
Africa	2,190
Middle East	7,994
Far East and Oceania	1,022
SUBTOTAL	11,389
Communist Nations	4,344
WORLD TOTAL	21,899

The primary exporter of oil is the Middle East which produced 36% of the world's petroleum in 1975. The Middle East produced 46% of the world's petroleum exclusive of that produced in Communist nations (which is not exported to W. Europe or Japan to any extent).

GROWTH OF OIL IMPORTS IN BILLIONS OF BARRELS PER YEAR

Year	*No. America*	*W. Europe*	*Japan*
1950	0.40	0.51	0.07
1960	0.80	1.31	0.22
1970	1.20	4.67	1.53
1975	2.15	4.79	1.84
1980 (Projected)	4.0	5.50	3.29

COST OF IMPORTS OF OIL TO THE U.S.

Year	*Cost (Billions of U.S. Dollars)*
1976	35
1977	45
1978	39.2
1979	52

The U.S. is importing over 40% of its oil, principally from the 13 nation Organization of Petroleum Exporting Countries (OPEC). OPEC includes not only the Middle Eastern nations of Saudi Arabia, Iran, Kuwait, Iraq, Qatar, and the United Arab Emirate, but also Venezuela, Ecuador, Indonesia, Nigeria, Gabon, Algeria and Libya. These countries hold four fifths of the proven crude reserves in the non-Communist world, and for some, oil provides over 90 percent of government revenues.

SUPPLIERS OF IMPORTED OIL TO THE U.S. IN 1978

Country	*Percent of Total*
Saudi Arabia	15.6%
Nigeria	11.9%
Iran	11.4%
Venezuela	10.5%
Libya	9.5%
Algeria	8.0%
Indonesia	7.4%
United Arab Emirates	5.9%
Canada	5.7%
Mexico	3.6%
Others	10.5%
TOTAL	100%

Supplies from Iran have been reduced in 1979 and Saudi Arabia has increased its production to account for the balance. About 83% of U.S. oil imports come from members of the Organization of Petroleum Exporting Countries (OPEC).

MIDDLE EAST CRUDE OIL PRICES

Period	*Posted Price*	*Arab Tax*
1968-69	$ 1.80	$0.88
Feb. to May 1971	2.18	1.26
Jan. to Mar. 1973	2.59	1.51
June 1973	2.90	1.70
Oct. 1, 1973	3.01	1.77
Oct. 16 - Dec. 31, 1973	5.12	3.05
Jan. 1, 1975	10.46	7.00
Jan. 1, 1976	11.51	7.00
Jan. 1, 1977	12.09	7.00
Jan. 1, 1978	12.70	7.00
Jan. 1, 1979	13.34	7.00
April 1, 1979	14.55	7.00
July 1, 1979	19.00	7.00
Jan. 1, 1980	24.00	7.00

ESTIMATED PROVED WORLD RESERVES OF PETROLEUM (IN 1974)

	Region (With Selected Countries)	Reserves (Billions of Barrels)	Production (Billions of Barrels)
I.	*North America*	*44.5*	*4.01*
	Canada	7.2	0.60
	U.S.	34.2	3.20
II.	*South America*	*21.1*	*1.48*
	Argentina	2.6	0.15
	Venezuela	14.6	1.09
III.	*Western Europe*	*20.1*	*0.11*
	Norway	5.5	0.01
	Great Britain	12.0	Negligible
IV.	*Eastern Europe*	*50.5*	*3.49*
	USSR	47.5	3.01
V.	*No. Africa*	*35.3*	*1.00*
	Libya	23.2	0.55
	Algeria	9.3	0.36
VI.	*Africa, Central & Southern*	*22.6*	*0.98*
	Nigeria	18.3	0.82
VII.	*Middle East*	*326.0*	*7.95*
	Iran	68.1	2.20
	Kuwait	70.9	0.83
	Saudi Arabia	96.9	3.00
	Iraq	35.7	0.74
VIII.	*Far East Asia*	*16.0*	*0.51*
	China (mainland)	14.8	0.26
	India	0.9	0.06
IX.	*Oceania*	*19.4*	*0.74*
	Indonesia	12.0	0.50
	Australia	2.5	0.14
	TOTAL WORLD	556.2	20.34

TOTAL DISCOVERIES OF OIL OVER 5 YEAR INTERVALS IN THE WORLD (EXCLUDING USSR AND CHINA)

	Average Annual Discovery Over 5-Year Period	*Average Annual Consumption Over 5-Year Interval*
1935-40	24	3.0
1940-45	8	4.0
1945-50	22	4.8
1950-55	18	5.5
1955-60	17	7.5
1960-65	18	10.0
1965-70	18	14.0
1970-75	18	18.0

WORLD PROVEN OIL RESERVES IN BILLIONS OF BARRELS

Year	*Proven Reserve*
1955	158
1960	290
1965	339
1970	531
1975	712

PRODUCTION AND PROVEN OIL RESERVES BY REGION OR NATION IN 1978

Region	*Reserves (Billions of Barrels)*	*Production (Millions of Barrels/Day)*	*Years Supply*
OPEC	420	31	40
USSR	70	11.0	17
U.S.	28	8.5	9
W. Europe	25	1.9	36
Mexico	27	1.7	59
China	18	2.0	27

Years supply is obtained by dividing reserves by 1978 production. *Source: Business Week,* July 30, 1979, pg. 45.

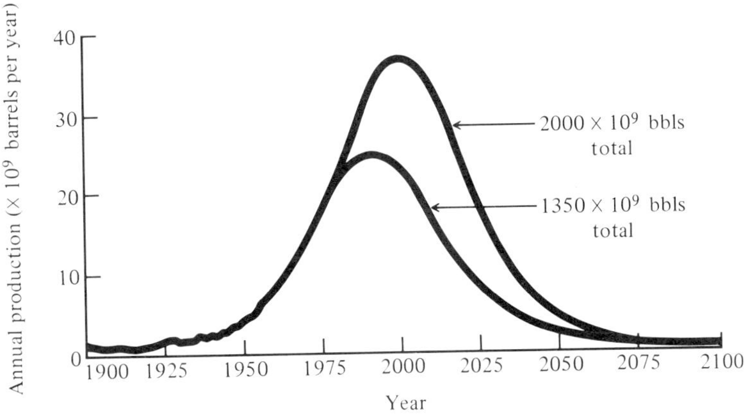

Petroleum production in the world for two estimated values of total oil resource

OIL WELLS DRILLED IN THE U.S.

Year	*Number of New Oil Wells Drilled*
1955	56,700
1960	46,700
1965	41,423
1970	29,500
1975	39,000
1978	48,000

Oil is discovered by small and large companies by means of geological exploration and the drilling of exploratory wells. Exploration is dependent upon economic factors, tax incentives and governmental regulations.

U.S. OFFSHORE OIL AND GAS RESOURCE ESTIMATES

	Offshore Oil (b.b.)	*Offshore Gas* ($\times 10^{12}$ *cu. ft.*)
Gulf of Mexico	30	220
Pacific	20	20
Atlantic	20	60
Alaska	30	190
TOTAL	100	490
	(Billions of Barrels)	(Trillions of Cubic Feet)

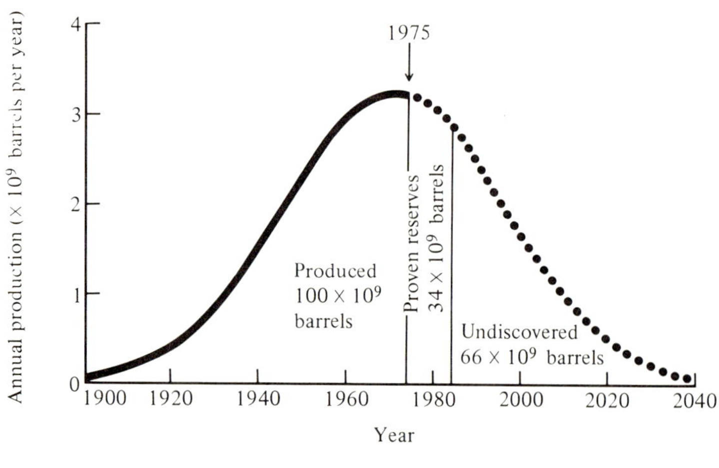

Petroleum production in the U.S. assuming a total resource of 200 billion barrels.

ANNUAL GASOLINE CONSUMPTION PER HOUSEHOLD FOR SELECTED STATES FOR 1979

State	*Consumption per Household (Gallons/Year)*
Wyoming	2,426
Texas	1,981
Georgia	1,853
Kansas	1,627
U.S. Average	1,498
California	1,435
Illinois	1,338
Massachusetts	1,233
New York	919

PETROLEUM PRODUCT INVENTORIES

Year	*Gasoline Inventories (Mar. 31 in Millions of Barrels)*	*Home Heating Oil Inventories (April in Millions of Barrels)*	*First Quarter Gasoline Consumption (Millions of Barrels per Day)*
1977	262	145	6.8
1978	260	139	6.9
1979	241	130	7.2

Refineries produce a balance of gasoline and home heating oil. Inventories have declined over the past several years.

U.S. SHALE OIL

Estimated Resource	1,000 Billion Barrels of Oil
U.S. Energy Plan Target for the Year 2000	500,000 Barrels/Day
Estimated Price of Production	$28/Barrel

Oil shale is a sedimentary rock which contains the solid hydrocarbon wax kerogen in tightly packed mud and clay. The Kerogen may be decomposed at elevated temperatures (450°C) resulting in an oil suitable for refinery processing. The primary obstacle to use of this oil shale is finding an economical, environmentally acceptable process for extracting the oil from the shale.

7 ELECTRICITY

Electricity serves as a carrier of energy to the user. Energy present in a fossil fuel or a nuclear fuel is converted to energy in the form of electricity in order to transport and readily distribute it to customers. By means of transmission lines, electric power is transmitted and distributed to essentially all the residences, industries, and commercial buildings in the U.S.

Electricity is a relatively useful energy carrier since it is readily transported with low attendant losses. Methods for safe handling of electricity have been developed over the past 80 years. Furthermore, methods of converting fossil fuels to electric power are well-developed, economical, and reliably safe. Means of converting solar and nuclear energy to electric power are currently in various stages of development or proven safety. Geothermal energy, tidal energy, and wind energy may also be converted to electric energy. The kinetic energy of falling water may also be readily used to generate electric power.

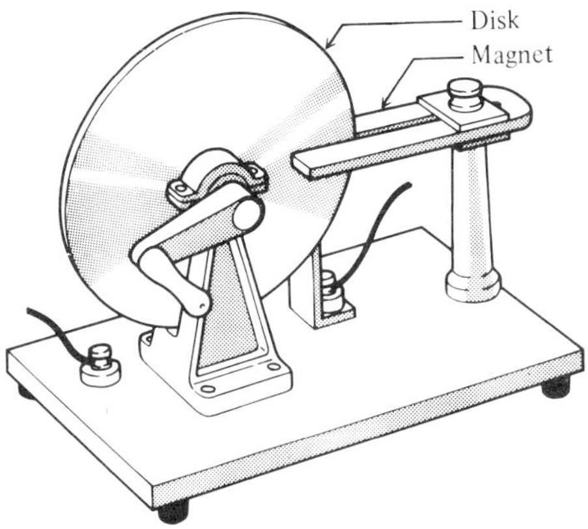

The first generator for producing electric current was built in 1831 by a young Englishman Michael Faraday. Faraday mounted a 12-inch metal disk on an axle between the poles of a magnet. When, by means of a crank, the disk was rotated through the magnetic field, an electric current was generated and flowed through the connecting wires.

ELECTRIC ENERGY USE IN THE U.S.

Year	*Population (Millions)*	*Generating Capacity (Million kW)*	*Consumption (Billion kWh)*	*Consumption per Persons (kWh)*
1900	76		3.7	49
1920	106		57.5	540
1930	123		115.0	935
1940	132		141.8	1,074
1947	144	52.3	255.7	1,774
1950	152	68.9	329.1	2,161
1955	166	114.5	547.0	3,297
1960	180.7	168.0	753.4	4,169
1965	184.2	236.1	1,055.0	5,433
1970	204.8	340.4	1,529.8	7,469
1971	207.0	367.5	1,614.0	7,800
1974	213.0	475.0	1,862.0	8,741
1978	218.0	500.0	2,206.0	10,120
1980 (Projected)	229.0	550.0	2,450.0	10,700
1985 (Projected)	236.0	630.0	3,000.0	12,710
1990 (Projected)	242.0	710.0	3,400.0	14,000

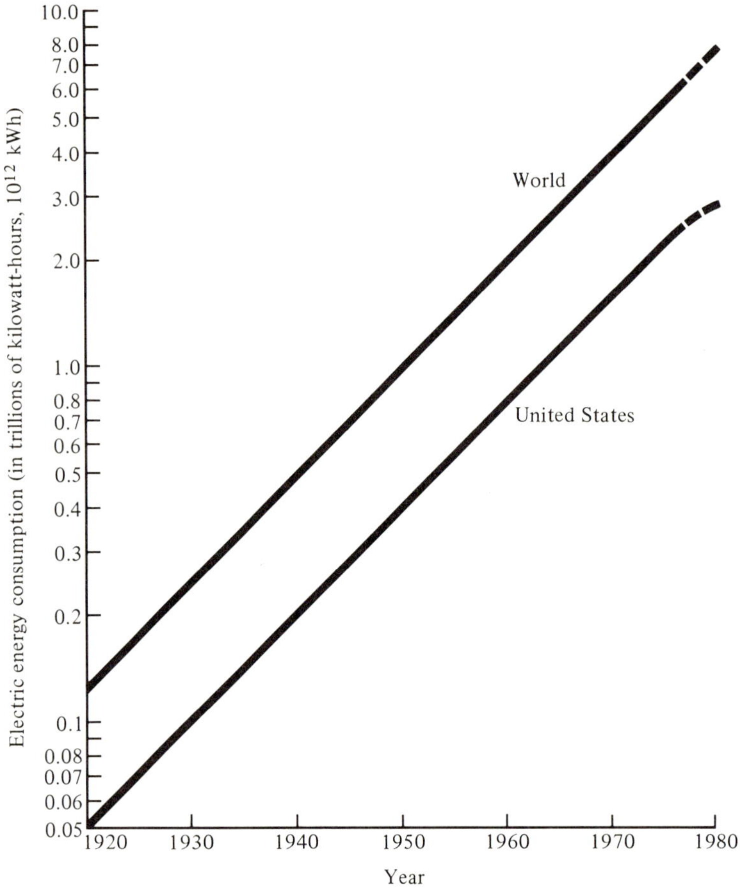

Electric power consumption in the United States and the world.

ELECTRICAL ENERGY CONSUMPTION PER CAPITA
(In Kilowatt-hours per Annum per Capita)

	USA	*UK*	*Japan*	*Hungary*	*Netherlands*	*Yugoslavia*	*Federal Republic Germany*
1930	935	280	230	100	210	60	
1940	1,074	620	470	200	420	50	
1950	2,160	1,120	550	320	730	150	860
1960	4,169	2,630	1,270	760	1,440	490	2,220
1970	7,500	4,440	3,100	1,460	3,300	1,220	4,000

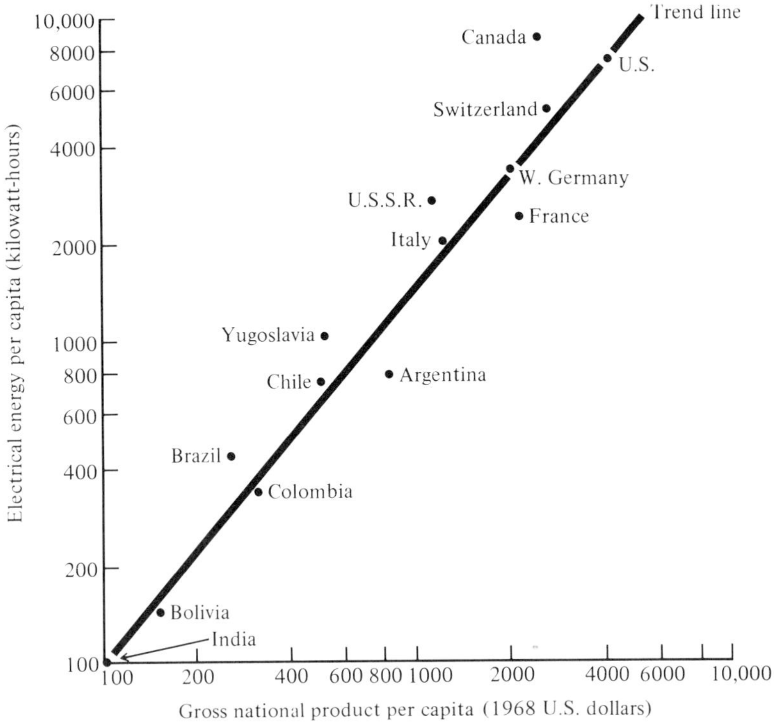

The electrical energy consumption per capita versus the Gross National Product per capita for several nations. (Note both scales are logarithmic.)

ELECTRICITY CONSUMPTION IN SEVERAL NATIONS IN 1971 IN BILLIONS OF KILOWATT-HOURS

	Consumption *(10^9 Kilowatt-Hours)*
USA	1,614
USSR	800
Japan	379
W. Germany	260
G. Britain	250
Canada	215
France	150
Yugoslavia	29
India	59
Brazil	48
Mexico	31

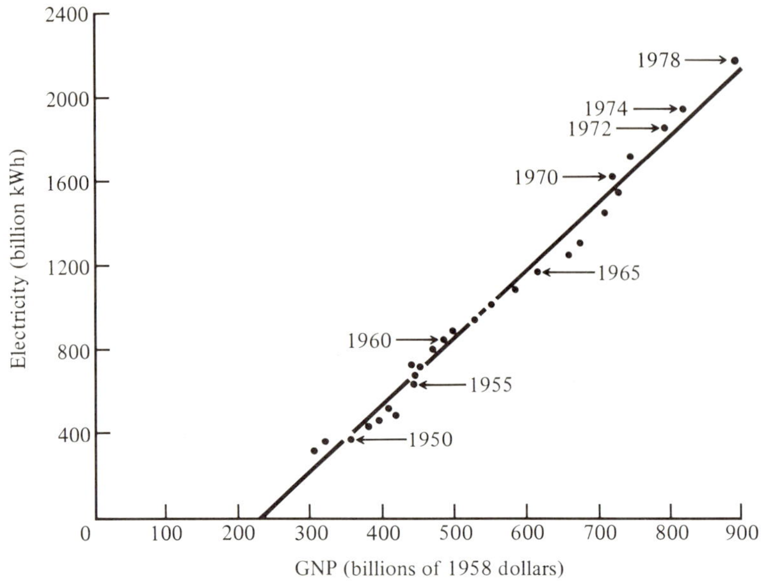

The growth of electricity consumption and the Gross National Product in the United States.

INDEX OF ENERGY USE AND ELECTRICITY USE FOR SELECTED NATIONS IN 1975

Country	*Index of Energy Use (Energy/GNP in \$)*	*The Percent of Total Energy Used to Generate Electricity*
USA	2.7	26
USSR	3.2	24
Canada	2.8	28
Japan	1.5	38
France	1.2	24
Sweden	1.6	42
W. Germany	1.6	28
United Kingdom	2.1	28
Mexico	1.8	16
Israel	1.0	34
World	2.4	22

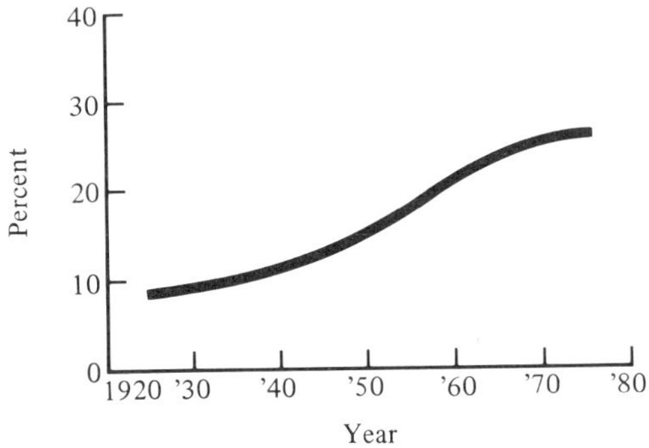

Percent of total energy converted to electricity in the United States.

STRUCTURE OF THE ELECTRIC UTILITY INDUSTRY IN THE U.S. IN 1970

Type of System	*Number of Systems*	*Number of Plants*
Investor owned	250	1,923
Federal	2	
Public Nonfederal	700	800
Cooperative	65	
	1,017	2,723

CLASSIFICATION OF CUSTOMERS IN 1973 IN THE U.S.

Type	*Number of Customers (Millions)*	*Energy Consumption (Billions of kWh)*	*Percent of Total*
Residential	67.84	608.4	35.5%
Commercial and Industrial	8.62	1,041.4	60.6%
Street and Highway Lighting	0.11	14.3	0.8%
Other Public Authorities	0.15	43.7	2.5%
Railroads and Railways		4.7	0.3%
Interdepartmental	4.77	5.0	0.3%
		1,717.8	100.0%

END USES FOR ELECTRICITY

End Use	*Percent of Total Electricity Use*
Industrial Drives	39.7
Refrigeration	11.6
Lighting	10.8
Water Heating	6.2
Electrolytic Processes	5.8
Air Conditioning	5.6
Space Heating	3.3
Direct Heat	2.6
Television	2.6
Cooking	2.1
Clothes Drying	1.0
Other (Includes Transportation, Computers, etc.)	8.7
TOTAL	100.0

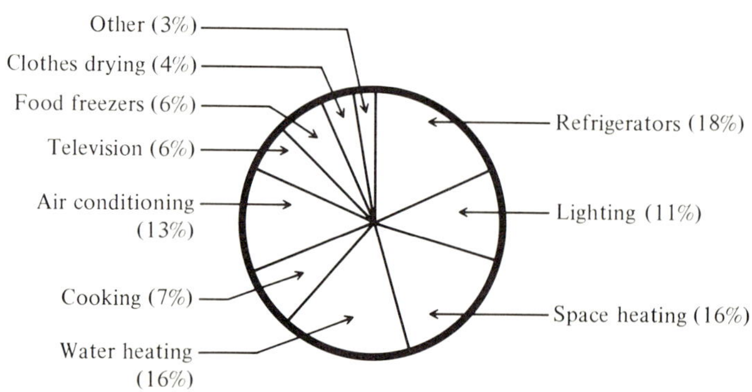

The uses of electricity in an average household in 1970 in the U.S. as a percentage of the total energy use. In 1970, 7,000 kilowatt-hours were consumed by the average household.

RESIDENTIAL USE OF ELECTRICITY IN THE U.S.

Year	*Customers (Millions)*	*Use per Customer (kWh)*	*Revenue per kWh (Cents)*	*Average Annual Bill (Dollars)*	*Residential Revenue (Million Dollars)*
1964	56.3	4,703	2.31¢	$108.64	6,041
1970	64.0	7,066	2.10¢	$148.39	9,416
1974	71.0	7,907	2.83¢	$223.77	15,703
1978	77.7	8,873	4.02¢	$357.07	27,400
1980 (Estimated)	80.3	9,300	4.14¢	$401.32	31,917

Source: Electrical World, March 15, 1979, pp. 74–75.

NET ELECTRICITY PRODUCTION BY PRIMARY ENERGY SOURCE IN 1977

Type of Power Plant	*Electricity Produced (Billions of kWh)*	*Percent of Total*
Coal	985.2	46.4
Petroleum	358.2	16.9
Natural Gas	305.5	14.4
Nuclear	250.9	11.8
Hydroelectric	220.5	10.4
Geothermal, Refuse, Wood and Other	4.1	0.2
TOTAL	2,124.4	100%

Two of the seven turbine generators at Moss Landing on Monterey Bay, California. The big pipes carry steam from the boiler to the turbine. Each generator shown has a capacity of 120,000 kilowatts. Two newer generators produce 750,000 kilowatts each. Moss Landing is the second largest power plant in the world. The boilers can burn either natural gas or petroleum. Photo courtesy of Pacific Gas and Electric Company.

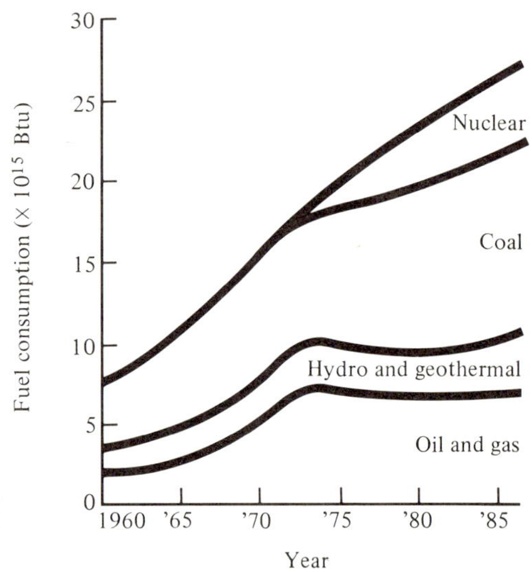

The actual and predicted energy sources used for the generation of electricity in the U.S.

GENERATING CAPABILITY NET ADDITIONS IN THE U.S. IN MEGAWATTS FOR SEVERAL SELECTED YEARS

Year	*Conventional Hydroelectric*	*Pumped Storage Hydro*	*Fossil Fuel Steam*	*Nuclear Steam*	*Combustion Turbine*	*Total*
1964	1,974		9,418	45	299	11,736
1970	1,789	313	16,800	2,513	6,126	27,741
1974	720	1,087	18,874	9,196	6,236	36,113
1978 (Projected)	899	2,031	13,855	10,140	1,006	27,931
1985 (Projected)	0	2,500	12,000	16,337	3,500	34,337

ANNUAL CAPITAL EXPENDITURES IN THE U.S. IN MILLIONS OF 1975 DOLLARS FOR THE PURPOSE OF GENERATING, TRANSMITTING AND DISTRIBUTING ELECTRIC ENERGY

Year	*Generation*	*Transmission*	*Distribution*	*Other*	*Total*
1964	1,819	1,056	1,800	267	4,940
1970	6,860	2,123	3,264	529	12,776
1974	12,504	2,450	4,576	1,024	20,555
1980 (Estimated)	10,410	2,064	5,570	902	18,946

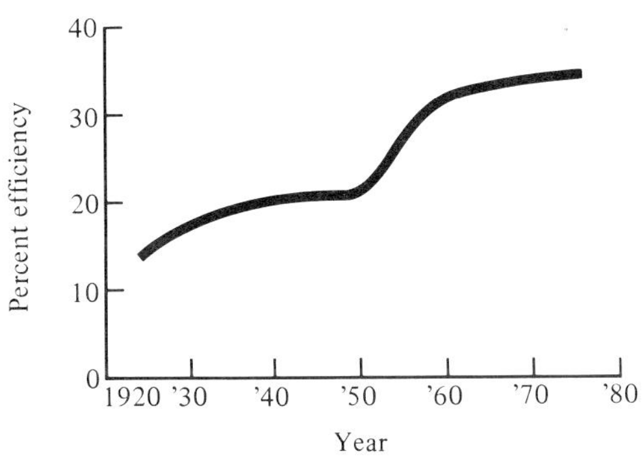

Overall efficiency of converting fossil fuel to electric energy for 1925 to 1975 in the U.S.

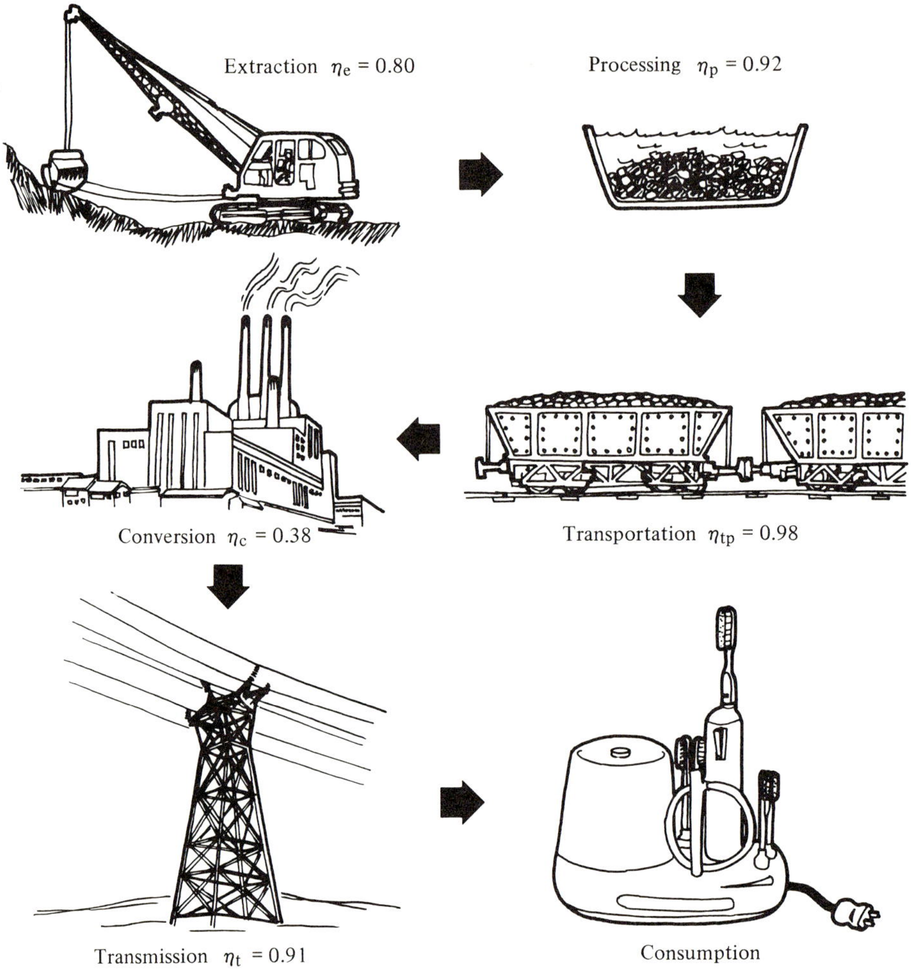

The movement of energy through the various stages of a coal-fired electric power system.

INCREASING ELECTRIC UTILITY BILLS FOR THE PERIOD 1972 TO 1979

Increase in Cost of Natural Gas	132%
Increase in Overall Consumer Prices	65%
Increase in Electric Rates to Residences	73%

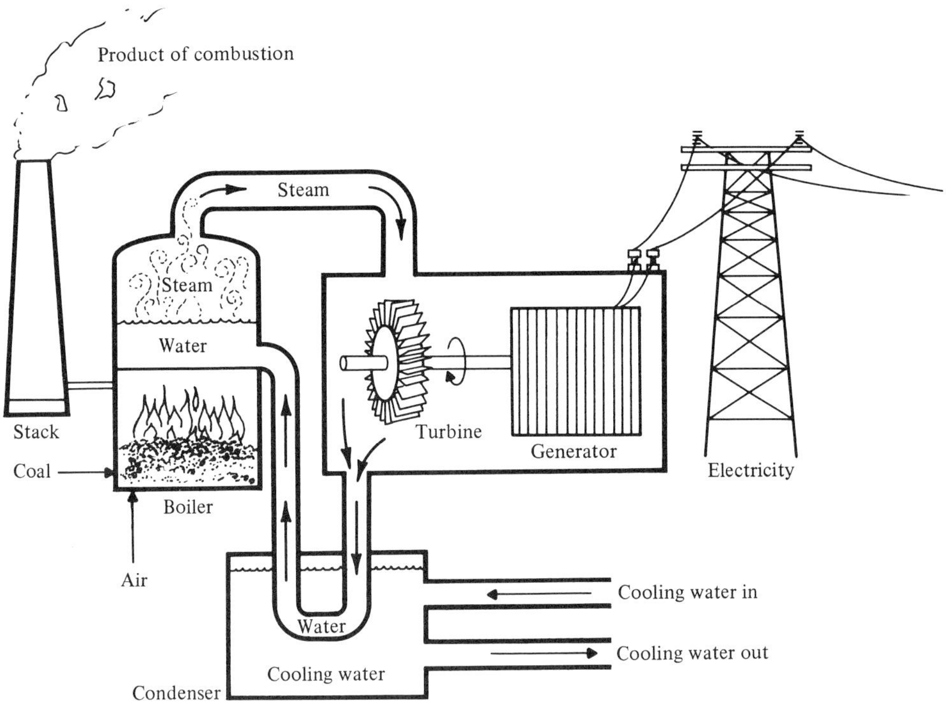

A diagram of a coal-fired electric power plant using a source of cooling water such as a lake or river.

THE AVERAGE ELECTRIC UTILITY BILL FOR 500 kWh FOR SELECTED CITIES IN 1979

New York City	$37.15
Cleveland	27.15
U.S. Average	23.76
Dallas	19.22
Portland	15.44
Seattle	8.89

The relatively low cost in Seattle and Portland is due to the existing extensive hydroelectric power available at low cost.

Source: U.S. News and World Report, April 30, 1979, pp. 45.

This tall tower supports wires carrying 500,000 volts of alternating current power. Voltages for transmitting electric power range from 100,000 volts up to 765,000 volts (alternating current). Courtesy of Pacific Gas and Electric Company.

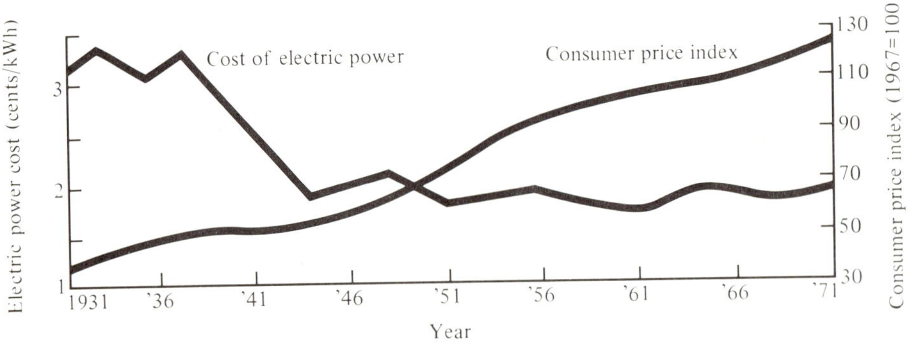

The average cost of electric power and the cost of living during the period 1931 to 1971.

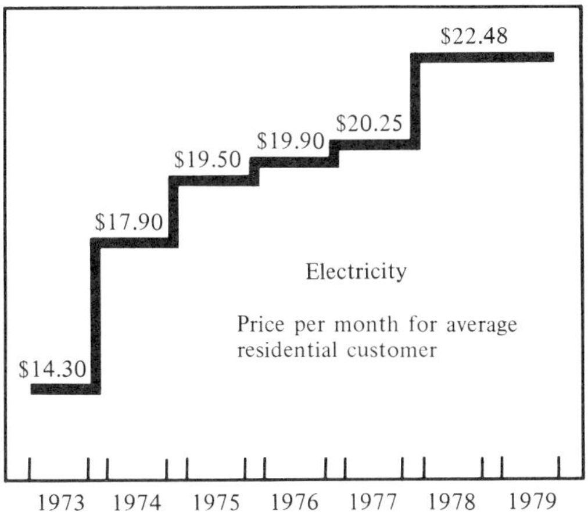

The price per month for an average residential customer for electric energy for the period 1973 to 1979 in Southern California. *Source:* Southern California Edison Company.

The Haynes Steam plant uses gas and oil for fuel. The 1625 megawatt plant contains six generators and was completed in 1967. Pacific Ocean water is used for condenser cooling. Today, a typical electric power plant (fossil fuel) can cost up to 500 million dollars to construct a 1,000 megawatt facility. Courtesy of Los Angeles Department of Water and Power.

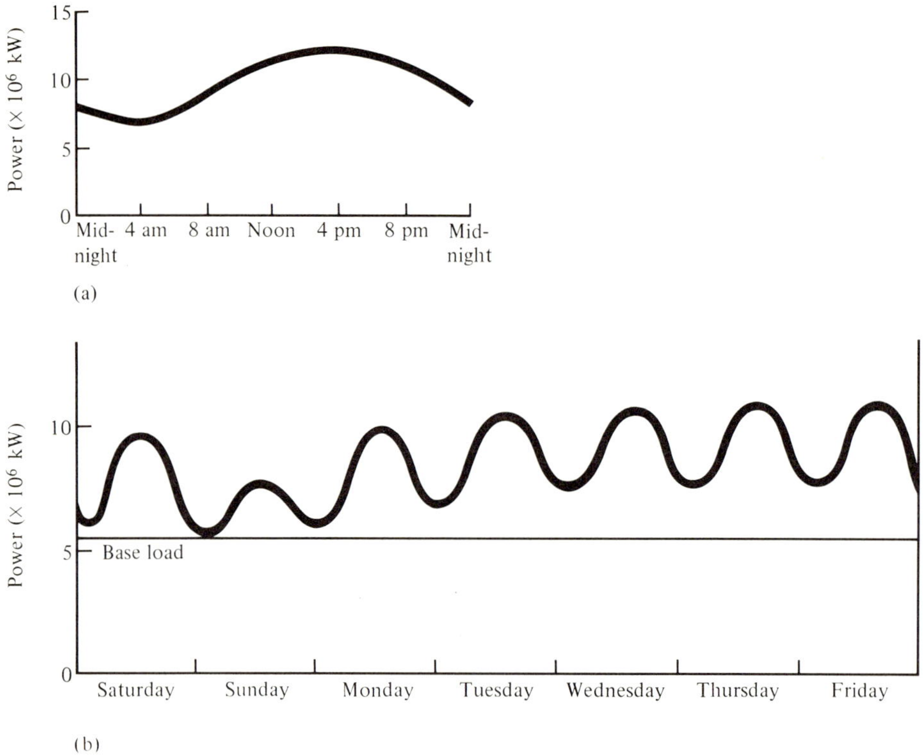

A typical load variation pattern for a day (a) and for a week (b).

FOSSIL FUELS AS PRIMARY ENERGY RESOURCES CONSUMED TO PRODUCE ELECTRICITY FOR 1972 TO 1978

Year	*Coal* (X 10^6 *Tons*)	*Petroleum* (X 10^6 *Barrels*)	*Natural Gas* (X 10^9 ft^3)
1972	352	491	3,976
1974	392	536	3,443
1976	448	556	3,081
1978	477	624	3,184

Source: Monthly Energy Review, U.S. Dept. of Energy, Feb. 1979.

COST COMPARISONS OF ALTERNATIVE POWER-GENERATING PLANTS, (1973 DOLLARS)

(Mills Per kWh at Generating Plant)

Item	Nuclear	Coal	Oil	Gas Turbine
	Type of Power-Generating Plant			
Fuel Cost	2.50	5.56	15.83	23.54
Operating and Maintenance Expenses[1]	1.50	2.75	2.65	2.50
Fixed Charges				
Load Factor = (.7)	14.84	11.48	10.45	5.12
Load Factor = (.4)	25.97	20.10	18.30	8.96
Total Costs				
Load Factor = (.7)	18.84	19.79	28.93	29.88
Load Factor = (.4)	29.97	28.41	36.78	32.76

One mil = 0.1¢ = 10^{-3} dollar

CHARACTERISTICS OF ELECTRICAL ENERGY SOURCES

Conventional Fossil-fuel Steam Electric

1. Air and Thermal Pollution (60% Heat Rejection)
2. Best Location Near Fuel Source and Cooling Water
3. Large World Reserves Available
4. Technology Well Developed and Proven
5. Relatively High Operating Cost

Hydroelectric Power Plants

1. Clean, Non-polluting and No Heat Rejection
2. Environmental Impact is Dam and Reservoir
3. Only Available Where Suitable Sites Exist
4. Largely Already Exploited in Industrialized Countries
5. Technology Well Developed and Proven
6. Very Small Operating Cost, While High Capital Cost per kW
7. 50 to 100 Year Economic Life

Nuclear Fission Reactors

1. Thermal Pollution (75% Heat Rejection)
2. High Capital Costs (Higher than Fossil Plants and Lower than Hydro)
3. Technology Available
4. Safety and Environmental Effects Not Proven

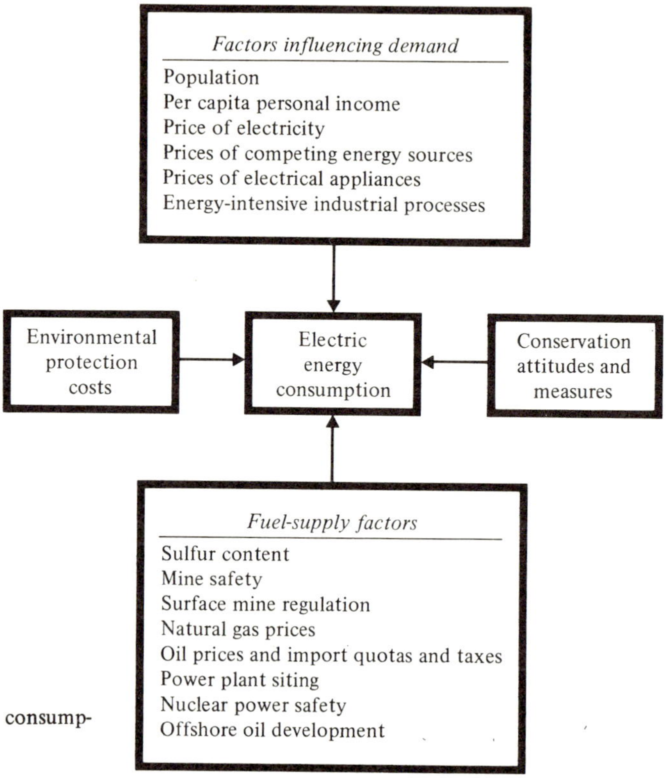

Interrelated factors influencing electric energy consumption.

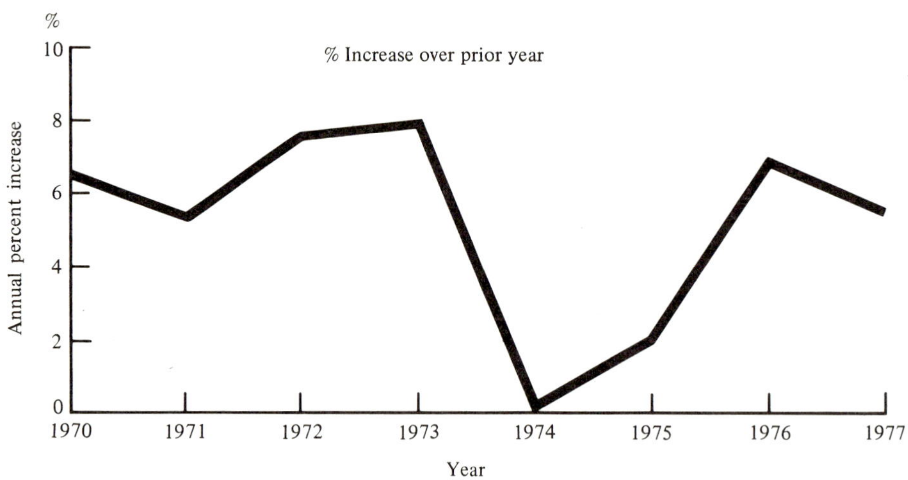

The annual percent increase has varied over the past decade for electric energy consumed in the U.S. (in kWh). A reasonable estimate for the future would call for annual increases of 4%, but conservation could reduce that increase.

8

TRANSPORTATION

The transportation of goods and people is integral to the needs of any society. Transportation technologies have evolved over the centuries, and now people may travel long distances at high rates of speed. Transportation systems use significant amounts of energy and are important to the whole social and economic fabric of any nation.

In the United States, transportation accounts for about 25 percent of the total energy use. About half of the petroleum consumed in the U.S. is used for autos, trucks, and airlines. Transportation is estimated to be responsible for about 20 percent of the gross national product (GNP).

Railroads were the principle form of transportation in the U.S. in the 1920s and have declined in use since then. Public mass-transit passengers use electric railway and motorbus, but the dominant means of personal transportation in the U.S. is the automobile.

RAILROAD USE IN THE U.S. IN THE 20TH CENTURY

Year	*Passengers (Millions)*	*Passenger-Miles (Millions)*
1900	577	16,038
1910	972	32,338
1920	1,270	47,370
1930	708	26,876
1940	456	23,816
1950	488	31,790
1960	327	21,284
1969	302	12,214
1975	271	9,582

PUBLIC TRANSIT PASSENGERS IN MILLIONS FOR THE PERIOD 1930 TO 1970 IN THE U.S.

Year	*Electric Railway*	*Trolley Coach*	*Motor Bus*
1930	13,000	16	2,479
1940	8,325	534	4,239
1950	6,168	1,658	9,420
1960	2,313	657	6,425
1970	2,116	182	5,034

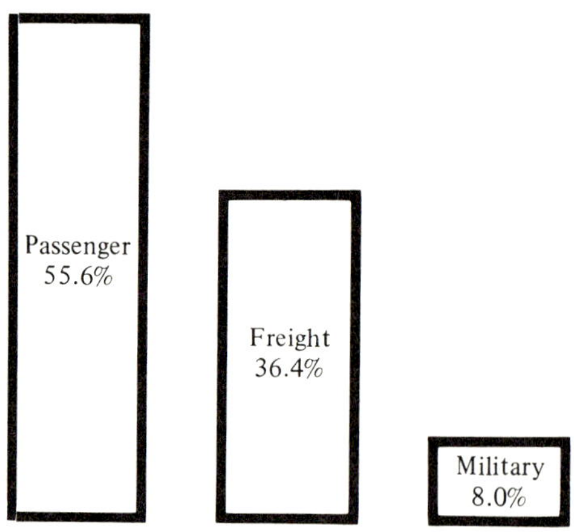

Distribution of energy consumption in 1971 in the U.S. in the transportation sector.

MOTOR FUEL USE AND MILES OF TRAVEL BY MOTOR VEHICLES FOR 1921 TO 1978

Year	*Highway Motor Fuel Use (Billions of Gallons)*	*Miles of Travel By Motor Vehicles (Billions of Vehicle-Miles)*	*Motor Vehicle Registrations (Millions)*
1921	3.9	55	8
1930	14.8	206	24
1940	22.0	302	29
1950	35.7	458	49
1960	57.9	719	65
1970	92.3	1,121	105
1972	103.2	1,250	115
1978	115.7	1,375	129

INTERCITY PASSENGER TRAFFIC IN THE U.S. IN BILLIONS OF PASSENGER-MILES AND THE PERCENT TRAVELLED ON EACH MODE

Year	*Total Volume (10^9 Passenger Miles)*	*Private Auto (Percent of Total)*	*Airlines (Percent)*	*Bus (Percent)*	*Railroad (Percent)*
1950	508	86.2	2.0	5.2	6.4
1955	716	89.0	3.2	3.6	4.0
1960	784	90.1	4.3	2.5	2.8
1965	920	88.9	6.3	2.6	1.9
1970	1.185	86.6	10.0	2.1	0.9

INTERCITY FREIGHT TRAFFIC IN THE U.S. IN BILLIONS OF TON-MILES AND THE PERCENT MOVED BY EACH MODE

Year	*Total Traffic ($\times 10^9$ Ton-Miles)*	*Railroads (Percent of Total)*	*Trucks (Percent)*	*Inland Waterways (Percent)*	*Oil Pipelines (Percent)*
1940	651	63	10	18	9
1950	1,094	57	16	15	12
1960	1,330	45	22	16	17
1970	1,921	40	22	15	22
1975		39	22	14	23

DEMAND FOR ENERGY INPUTS IN THE TRANSPORTATION SECTOR IN THE U.S. (IN 10^{12} Btu)

Year	*Coal*	*Petroleum*	*Natural Gas*	*Electric Energy*	*Total*
1950	1,701	6,785	130	24	8,640
1960	87	10,372	360	18	10,836
1970	8	15,592	745	16	16,361

THE TREND OF TOTAL ENERGY CONSUMPTION IN THE U.S. FOR THE PERIOD 1955 TO 1971 FOR SEVERAL MODES OF TRANSPORTATION

Mode	*1971 Consumption (Trillion Btu)*	*1955-1971 Average Annual Rate of Growth*	*1971 Percent of Total*
Automobiles	8,652	4.6	48.2
Trucks	3,445	4.7	19.2
Pipelines	1,590	6.0	8.9
Military	1,443	2.3	8.0
Airlines	698	15.2	7.1
Water	544	–0.9	3.9
Railroad	123	–2.7	3.0
Buses	89	1.3	0.7

THE ENERGY INTENSIVENESS OF VARIOUS MODES OF TRANSPORTATION FOR PEOPLE

Mode	*Btu/Passenger Mile*	*Total Percent of Transportation Demand*
Airplane	7,150	9.3
Automobile	5,400	88.4
Train	2,620	0.7
Bus	1,700	1.2
Walking (3 mph)	524	less than 0.1
Bicycle (8 mph)	310	less than 0.1

Energy Intensiveness: The energy required to move a passenger or freight a given distance.

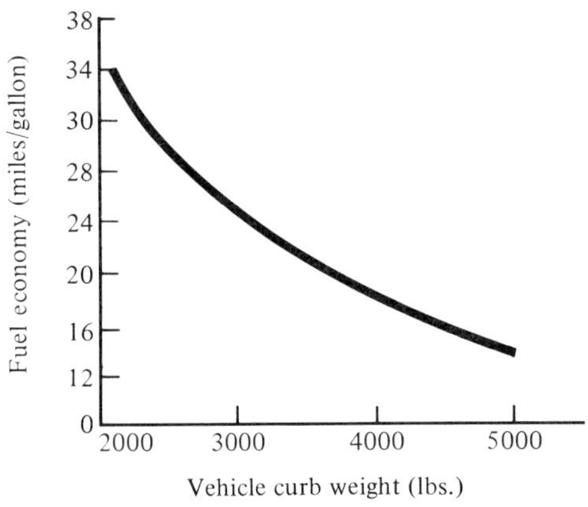

The fuel economy of 1978 U.S. automobiles as a function of the vehicle weight.

THE ENERGY INTENSIVENESS OF VARIOUS MODES OF TRANSPORTATION FOR FREIGHT

Mode	*Btu/Ton-Mile*
Pipelines	450
Waterways	540
Railroads	680
Trucks	2,340
Aircraft	37,000

EFFECTS ON THE FUEL ECONOMY OF THE AUTOMOBILE

Item	*Normal Effect*
Air Conditioner	Reduces Fuel Economy 9%
Automatic Transmission	Reduces Fuel Economy 6%
Steel Belted Radial Tires	Increase Fuel Economy 10%
Reducing Driving Speed from 65 mph to 55 mph	Increases Fuel Economy 5%
Reduce Weight of Vehicle from 4,000 to 3,500 lbs.	Increases Fuel Economy 15%

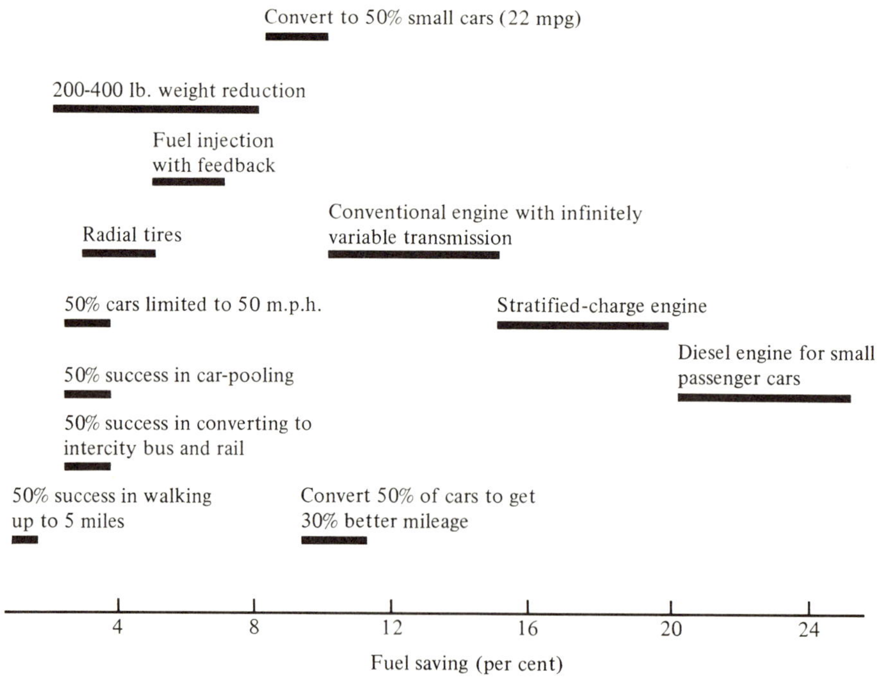

Various options proposed for saving petroleum used as automobile fuel. *Source:* M. Chiusano, "Conserving Transport Energy", *Technology Review,* June 1974, pp. 54–55.

U.S. DOMESTIC CONSUMPTION OF GASOLINE

(Average Daily Consumption in Million Barrels/Day)

1963	4.479
1965	4.713
1967	5.048
1970	5.839
1973	6.674
1975	6.6675
1977	7.176
1978	7.403

Source: Monthly Energy Review, U.S. Dept. of Energy, Feb. 1979.

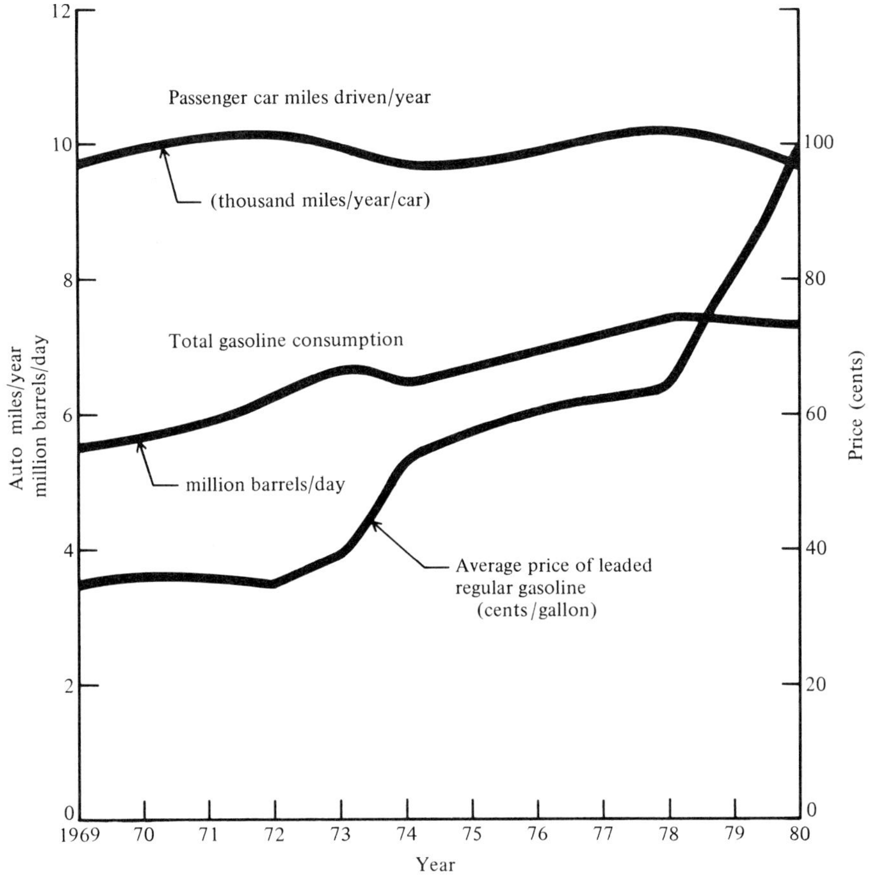

Increasing gasoline prices have leveled gasoline consumption and passenger miles driven per auto per year. *Source:* J. M. Berry, "Carter's Energy Policy" *Fortune,* Aug. 13, 1979, pp. 106–109.

THE ELECTRIC AUTOMOBILE

Characteristics of Electric Automobiles

Advantages	*Disadvantages*
Quiet Motor	Limited Range of Batteries
Easy to Operate	Current High Cost of Batteries
No Pollution (At the Auto)	Limited Top Speed
Smooth Acceleration	

The Copper Electric Town Car built by the Copper Development Association. The car has a range of 120 miles at a cruising speed of 40 mph and a top speed of 55 mph. It has a range of 75 miles in stop and go city driving and energy costs are estimated between 2¢ and 3¢ per mile. The car weighs 2,952 pounds including the 1,200 pounds of 18 batteries. The front wheel drive uses a separately excited motor. Photo courtesy of Copper Development Association.

DESIRED CHARACTERISTICS OF AN ELECTRIC VEHICLE

Minimum Urban Range	75 Miles
Maximum Electric Recharge Energy	0.5 kWh/Mile
Maximum Recharge Time	6 Hours
Minimum Top Speed	60 mph

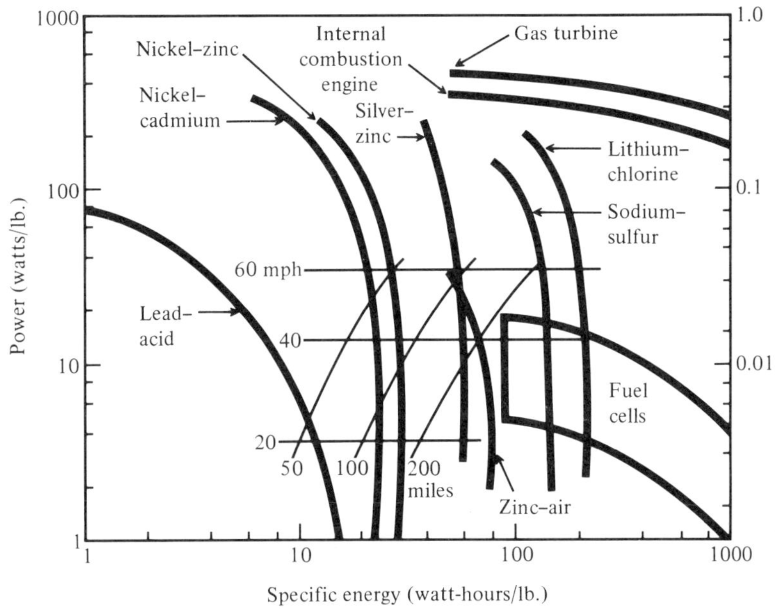

Vehicle requirements and motive power sources for an electric car.

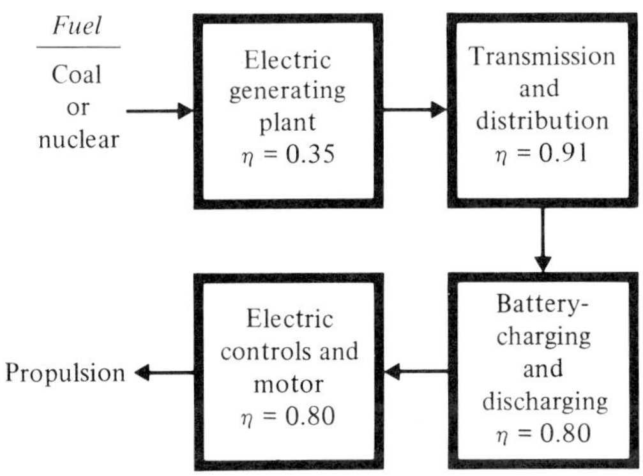

A schematic diagram of the propulsion efficiency of an electric car. The overall efficiency is 20%.

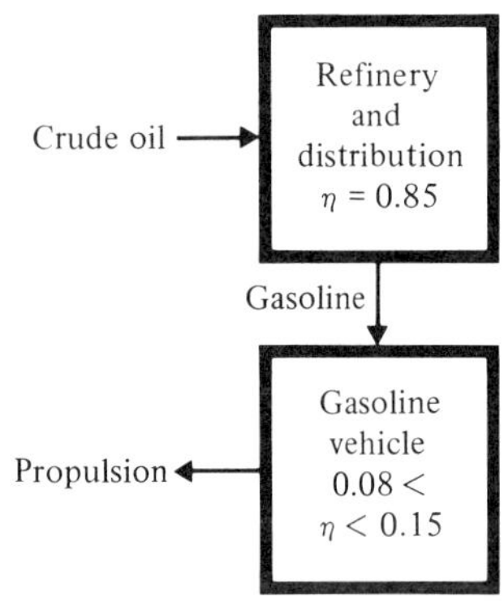

A schematic diagram of the propulsion efficiency of a gasoline powered car. The overall efficiency is about 10%.

The Electric Test Vehicle One, developed for the Department of Energy, was first demonstrated on June 22, 1979. The life cycle cost is estimated to be 18 cents per mile. The 3,300 lb. car used 18 lead-acid batteries and provides a 100 mile range at up to 45 mph. The car was developed by General Electric Company with Chrysler Corporation as a major subcontractor.

9
AGRICULTURE

The objective of agriculture is to collect and store solar energy as food energy in plant and animal products. These products are then distributed throughout the world to serve as food for the population. To collect solar energy in plants and animals, farmers must spend fossil-fuel energy and electric energy in tillage, fertilizers, irrigation, harvesting, and processing to enable crops to convert solar energy into food energy.

Energy has helped to make a revolution in farm life, farm work, and farm output since 1900. Only 100 years ago in the U.S. we were a rural people: farmer, planter, trapper, and pioneer. In the 19th century most work was done on farms by musclepower—human or animal. The chief sources of energy used for heating were wood and coal. Most farms were fortunate enough to have a windmill for pumping water. In 1900, one farmworker was able to supply the needs of about seven people. Today, a farmworker supplies the needs of 50 people.

THE NUMBER OF FARMS, LAND IN FARMS AND THE FARM POPULATION FOR 1900-1977

Year	*Number of Farms (Thousands)*	*Land in Farms (Million Acre)*	*Average Acres per Farm*	*Farm Population*	
				Millions of People	*Percent of Total Population*
1900	5,737	839	146	29.4	38.7
1910	6,406	879	137	32.2	34.9
1920	6,518	956	147	32.0	30.1
1930	6,546	987	151	30.5	24.9
1940	6,350	1,061	167	30.5	23.2
1950	5,648	1,202	213	23.0	15.3
1959	4,105	1,183	288	16.6	9.4
1969	2,971	1,124	378	10.3	5.1
1971	2,876	1,117	389	9.4	4.6
1977	2,700	1,072	397	7.8	3.6

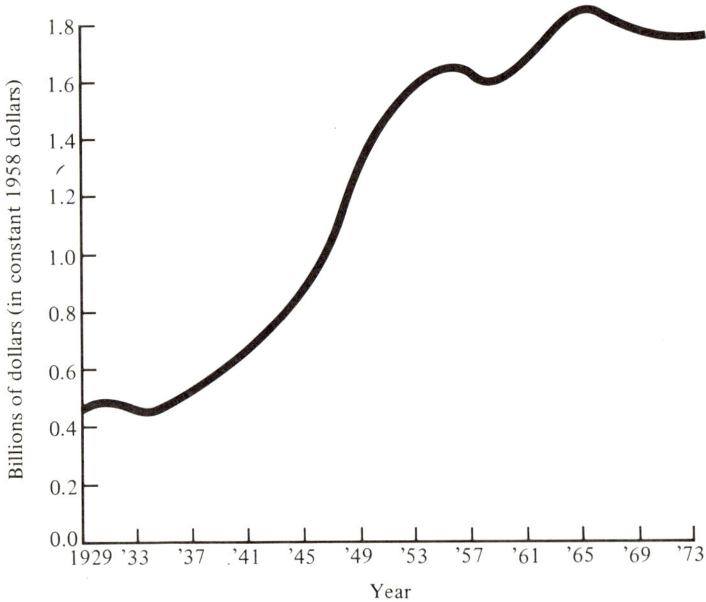

Annual farm energy expenditure in the U.S. for the period 1929–73.

USE OF ENERGY INTENSIVE PRODUCTS IN U.S. AGRICULTURE

Year	*Tractors on Farms (Thousands)*	*Commercial Fertilizer Used (Thousands of Tons)*	*Value of Farm Implements and Machinery (Millions of Constant 1967 Dollars)*
1900	–	2,730	2,727
1910	1	5,547	3,543
1925	549	7,329	5,544
1940	1,545	8,336	7,345
1950	3,394	20,345	14,873
1970	4,619	39,775	28,928

AGRICULTURAL OUTPUT AND PRODUCTIVITY IN RELATION TO INPUT RESOURCES FOR 1929-72

Variables in Constant 1958 Dollars	*1929*	*1948*	*1958*	*1972*	*Annual Percent Change 1929-1972*
Gross Farm Product (Billion Dollars)	17.0	19.0	20.8	24.7	1.0
Hired and Family Labor (Billion Hours)	23.2	16.8	10.5	6.5	–3.0
Farm Energy Expenditures (Billion Dollars)	0.5	1.2	1.6	1.7	4.09
U.S. Population (Millions)	121.8	146.6	174.9	208.8	1.39
Per Capita Income (Dollars)	1236	1567	1831	2771	2.27
Agricultural Exports (Billion Dollars)	2.6	2.2	4.0	6.9	3.75
Persons Supplied by One Farm Worker	9.5	15.0	24.8	49.2	
Index of U.S. Crop Production per acre (1967 = 100)	57	67	80	120	

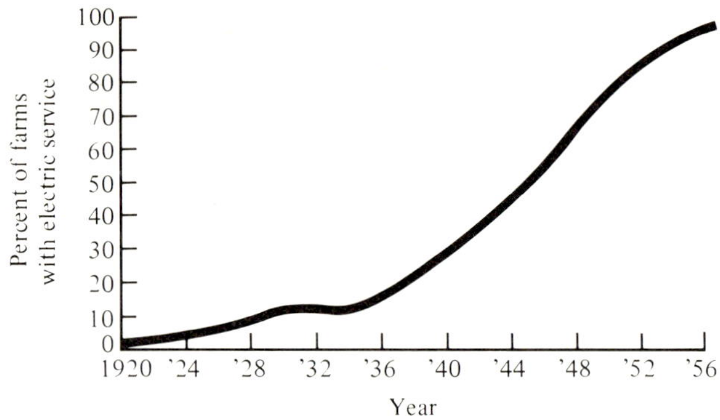

The growth of electric service to farms in the U.S. during the period 1920 to 1956.

ENERGY USE IN THE U.S. FOOD SYSTEM
VALUES GIVEN IN 10^{12} KILOCALORIES

	1950	*1970*
The Farm		
Fuel	158.0	232.0
Electricity	32.9	63.8
Fertilizer	24.0	94.0
Agricultural Steel	2.7	2.0
Machinery Manufacture	30.0	80.0
Tractors Manufacture	30.8	19.3
Irrigation	25.0	35.0
SUBTOTAL	303.4	526.1
Food Processing Industry		
Processing Industry and Machinery	197.0	314.0
Packaging and Containers	105.0	207.0
Transport Fuel and Transport Manufacture	151.5	320.9
SUBTOTAL	453.5	841.9
Commercial and Home		
Commercial Refrigeration and Cooking	150.0	263.0
Refrigeration Machinery	25.0	61.0
Home Refrigeration and Cooking	202.3	480.0
SUBTOTAL	377.3	804.0
GRAND TOTAL	1,134.2	2,172.0

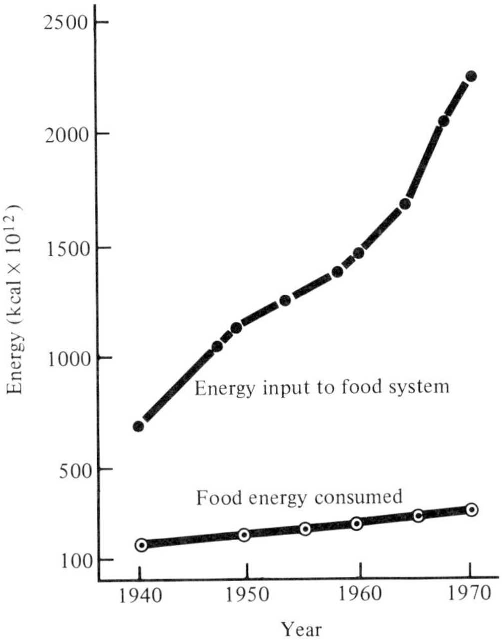

The energy use in the U.S. food system for the period 1940 to 1970 and the annual caloric content of the food consumed during each year of that period.

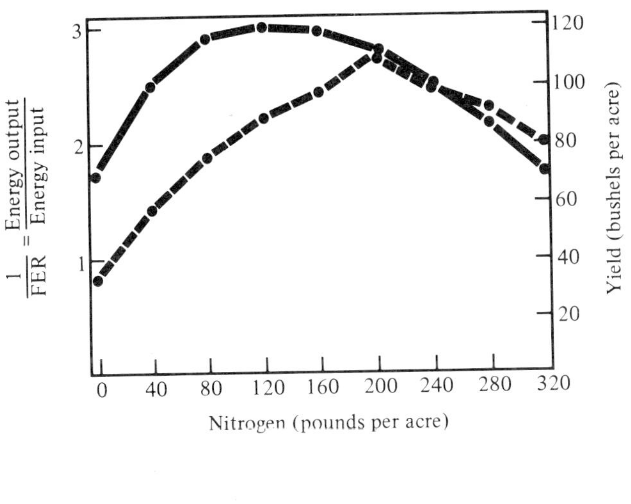

The energy output obtained divided by the energy input for corn versus the amount of nitrogen fertilizer applied (solid line). The dotted line shows the yield per acre of corn versus the nitrogen fertilizer applied. The maximum ratio of output to input energy of 3.0 is obtained at about 120 lbs. of nitrogen per acre.

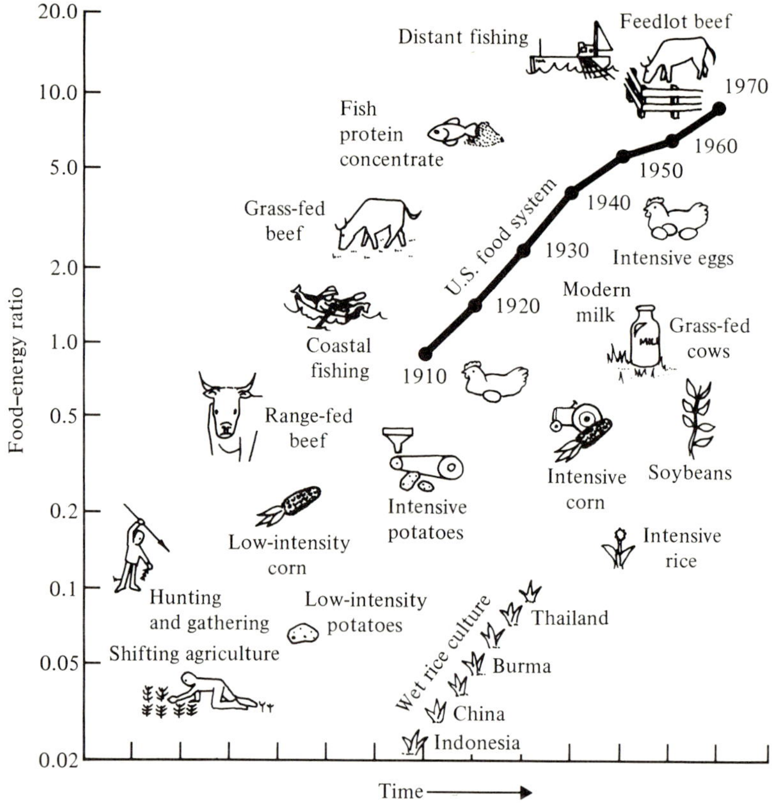

The food energy ratio, which is the input energy required per energy of food consumed for various food crops. Agricultural methods evolve over the centuries towards the right side of the graph. *Source:* J. S. Steinhart and C. E. Steinhart, "Energy Use in the U.S. Food System," *Science,* April 19, 1974, pp. 307–315.

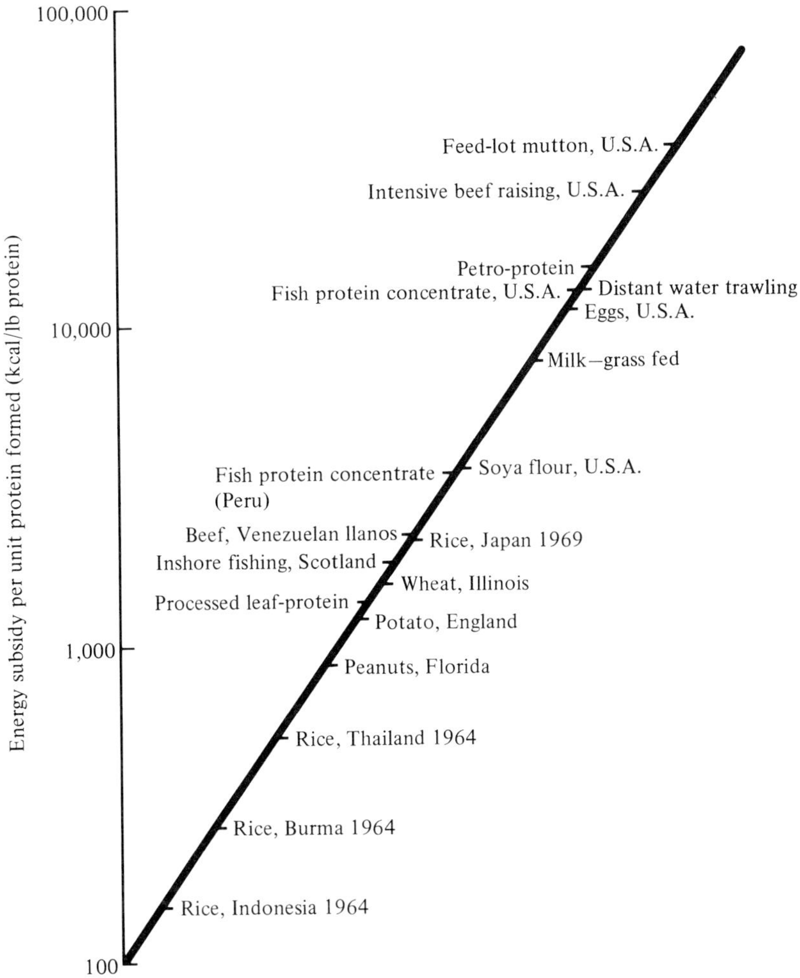

The difference between the input energy and the output energy is defined as an energy subsidy. The energy subsidy per unit of protein obtained for several sources of protein. *Source:* M. Slesser, "Energy Analysis in Technology Assessment," *Technology Assessment,* Vol. 2, No. 3, 1974, pp. 201–208.

10

HYDROELECTRIC POWER GENERATION

Energy flows in the hydrologic cycle of evaporation, precipitation, and surface runoff of water and can be harnessed to generate electric power. World-wide, this flow is 4×10^{16} watts, and only a small fraction is used by man to turn water turbines and electric generators.

The earliest use of water power was in water mills several thousands of years ago. By the 16th century water mills had to be adapted to many processes. The use of water falling through a distance in order to turn a turbine and then an electric generator began after the general introduction of electric power in the U.S. and Europe at the end of the 19th century. In 1882 a hydroelectric station was placed in operation on the Fox River in Appleton, Wisconsin. A waterwheel drove two Edison-type generators with a total output of 25 kilowatts. This was sold to two paper mills and one residential customer for lighting incandescent lamps. In 1895 the first large hydroelectric power plant was built at Niagara Falls, New York.

Generators in the first Niagara Falls power station, Photo 1896. (Courtesy of the Niagara Mohawk Power Corporation.)

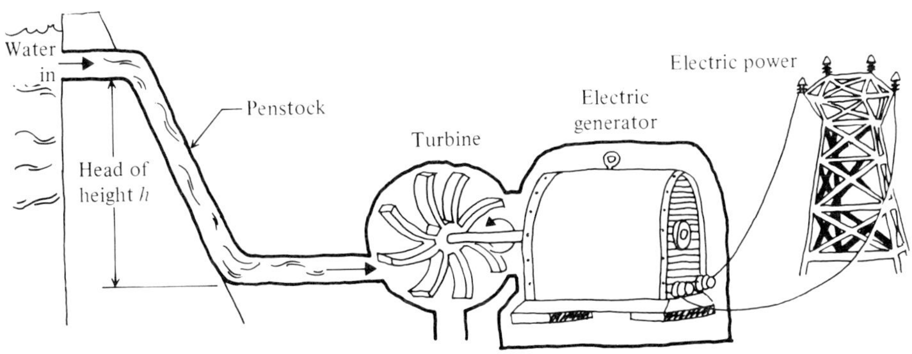

Schematic diagram of a hydroelectric generating plant. The water falls through a head driving the turbine which rotates the electric generator.

A hydroelectric power plant using a head of 1193 feet on the Pit River in California. The water decends from a reservoir on the McCloud River through 5500 feet of steel penstock. (Courtesy of Pacific Gas and Electric Company.)

A Pelton turbine runner. This photograph, looking almost directly up from the pit, shows the fabricated bucket wheels of a vertical impulse type turbine and some of the nozzles and jet deflectors. This 39,000 hp (29-megawatts) six jet installation operates at 429 revolutions/minute under a 1233-foot head. (Courtesy of Allis-Chalmers Corporation.)

A Francis-type turbine runner with a diameter of 19 feet. This type of turbine is used with heads of 100 to 1,000 feet. This runner is used in a 140,000-horsepower (104-megawatt) turbine operating at 100 revolutions/minute under a 162-foot head. (Courtesy of Allis-Chalmers Corporation.)

WATER POWER IN THE UNITED STATES

Year	*Capacity (Megawatts)*	*Electric Energy Output Provided (Gigawatt-hours)*
1920		18,779
1925		25,496
1930		35,878
1935		42,727
1940		50,131
1945	15,892	84,747
1950	18,675	100,885
1955	25,742	116,236
1960	33,180	149,423
1965	42,948	196,984
1970	51,952	250,611
1971	53,404	269,580
1972	53,778	272,600
1975	60,000	289,000
1979	74,000	307,000
1984*	84,000	340,000
Total Potential = 200,000		

*Projected

INSTALLED HYDROPOWER CAPACITY OF THE WORLD IN 1969

Region	*1969 Installed Hydroelectric Capacity,* 10^3 Mw_e	*Percentage of Total Installed Electrical Generating Capacity*
North America	83.9	22.1
Central America	0.5	41.7
West Indies	0.2	4.5
South America	13.4	50.0
Europe, Including U.S.S.R.	126.4	26.4
Africa	6.1	29.7
Asia	37.6	32.1
Oceania	6.6	35.4
WORLD	274.7	26.3

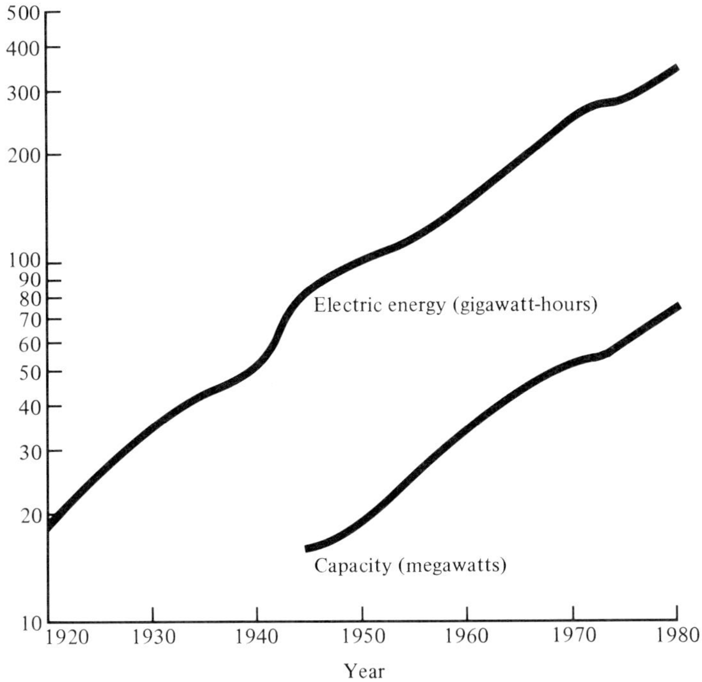

Electric energy and capacity provided by hydroelectric generating stations in the period 1920 to 1980 in the U.S.

The Hoover Dam and Electric Power Plant was completed in 1936. The dam is 726 feet high and 660 feet thick at the bottom. Lake Mead behind the dam is 110 miles long. The generating plant has 17 generators and a capacity of 1345 megawatts. (Courtesy of Los Angeles Department of Water and Power.)

ESTIMATED POTENTIAL HYDROPOWER CAPACITY OF REGIONS OF THE WORLD

Region	*Potential Hydropower (10^3 Megawatts)*	*Percent Developed*
Africa	437	10%
Asia (Less USSR)	684	5%
Europe (Less USSR)	215	50%
USSR	269	10%
North America	330	30%
South America	288	8%
Oceania	37	20%

Water can be pumped up to a reservoir by electric pumps and then allowed to fall through a turbine and generate electricity when it is needed. The Lundington pumped storage project in Michigan can yield 1.9 gigawatts maximum power and 15,000 megawatt-hours of stored energy. The Lundington plant uses Lake Michigan as the lower reservoir while the upper reservoir is a man-made lake measuring about 3 km by 1.6 km. It cost $340 million to construct. The upper reservoir is 105 meters above Lake Michigan. The Lundington facility is the largest pumped storage project in the world. (Photo courtesy of Ebasco Services Incorporated.)

POWER CAPACITY OF SELECTED HYDROELECTRIC FACILITIES

Name and Location	*Capacity (Megawatts)*
Hoover Dam, Nevada	1,345
Grand Coulee Dam, Washington	4,200
Aswan High Dam, Egypt	2,100
Churchill Falls, Canada	5,225
Bratsk Station, Siberia, USSR	4,150

COMPARISON OF ADVERSE EFFECTS ON SEVEN ENVIRONMENTAL PARAMETERS OF HYDRO, NUCLEAR AND FOSSIL FUEL POWER PLANTS

Adverse Effects	*Hydro*	*Nuclear*	*Fossil Fuel Steam*
Land Use	Greatest Effect	Great Effect	Least Effect
Major River Regulation	Yes	No	No
Thermal Pollution	No	Yes	Yes
Air Pollution	No	Yes	Yes
Radioactive Waste	No	Yes	No
Consumes Natural Resources	No	Yes	Yes
Sensitivity to Earthquakes	Moderate	Greatest	Least

The Pit 7 powerhouse of Pacific Gas and Electric on the Pit River, California, has two Francis (reaction-type) turbines on vertical shafts utilizing a large volume of water under a low head. (Courtesy of Pacific Gas and Electric Company.)

11

GEOTHERMAL ENERGY

Over the life of the earth, heat energy has been stored within the earth's core. Of this heat stored within the earth, some is conducted from the interior by means of conduction through the earth. Unfortunately, we cannot directly exploit this heat supply, but we can use local hot spots—subterranean reservoirs where the heat has been stored in the form of steam and hot water. Such reservoirs are the source of geothermal energy.

Geothermal reservoirs consist of permeable and porous rock in which, by circulation of steam or hot water, a convection system can develop. The porosity and permeability need not be inherent in the type of rock itself, but can result from fracturing due to deformation. Ground water, which can percolate down to depths of several miles, is heated directly or indirectly by the underlying magmas, expands, and ascends towards the surface. A geothermal reservoir can store enormous amounts of energy in the form of steam and hot water.

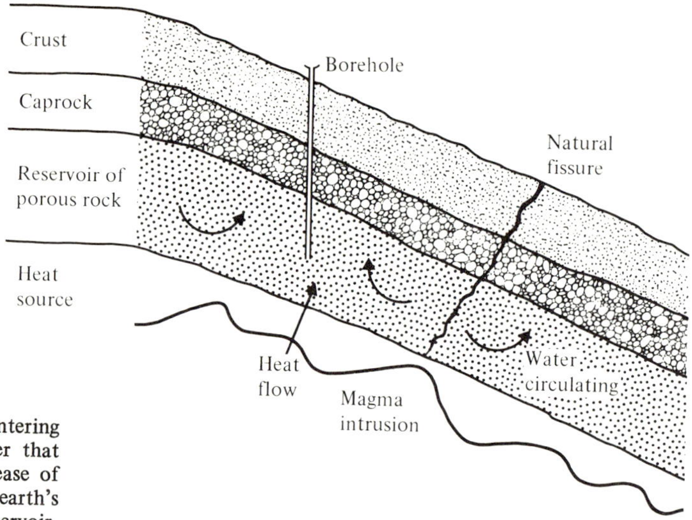

A diagram of a geothermal reservoir. Water entering the reservoir may come from groundwater that seeps down from the surface or from release of water from the molten rock (magma) of the earth's core. Molten rock heats the water in the reservoir. Caprock traps the heat of the water until it is released by a well or a natural fracture.

The Geysers Power Plant in Sonoma County north of San Francisco, California, is the only geothermal plant active in the U.S. The geysers are located on the steep slopes of Cobb Mountain, an extinct volcano. Steam is obtained from fumaroles and steam vents which emit steady steam vapor, not intermittent geysers despite the name of the site. These two units shown went into operation in 1960 and 1963 respectively and have a combined capacity of 24 megawatts. The capacity of the site in 1979 was 600 megawatts. Courtesy of Pacific Gas and Electric Company.

ELECTRIC POWER CAPACITY OF GEOTHERMAL PLANTS IN 1975

Larderello, Italy	406 Megawatts
Wairakei, New Zealand	145 Megawatts
The Geysers, California	516 Megawatts
Matsukawa, Japan	30 Megawatts
Cerro Prieto, Mexico	75 Megawatts
WORLD TOTAL	1,172 Megawatts

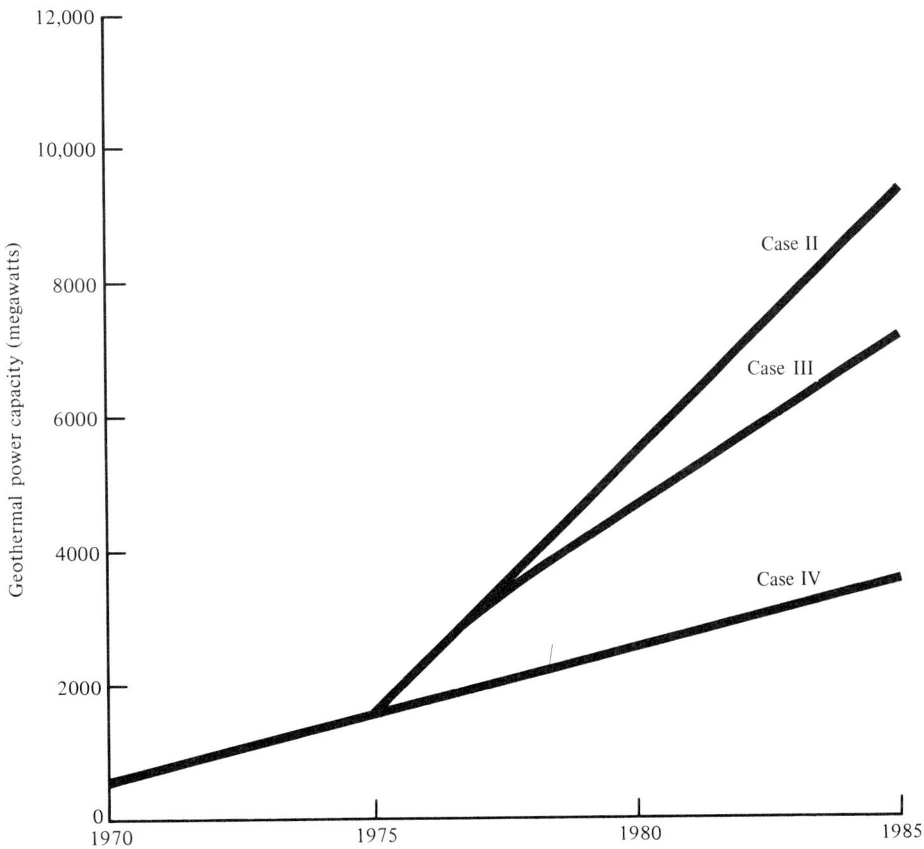

Actual and projected capacity of geothermal power plants in the U.S. for three possible rates of growth for all uses. *Source: Energy Facts,* U.S. Gov. Printing Office, 1973.

CHARACTERISTICS OF FOUR TYPES OF GEOTHERMAL SOURCES

Type/Characteristic	*Dry Steam*	*Wet Steam*	*Hot Brine*	*Hot Rock*
Ease of Utilization	Highest (Greatest Ease)	Somewhat Difficult	Difficult	Greatest Difficulty
Abundance	Lowest	Somewhat Abundant	Abundant	Great Abundance
Drilling Depth	8,000 Ft. (2,440m)	3,000 Ft. (915m)	4,500 Ft. (1,370m)	10,000 Ft. (3,050m)
Pressure and Temperature of Steam or Water	7 Atmospheres 200°C	2 Atmospheres 230°C	1 Atmosphere 240°C	400°C
Currently Operating Fields	Larderello, Italy Geysers, USA Matsukawa, Japan	Wairakei, New Zealand Cerro Prieto, Mexico	None	None
Exploration Sites		Kilauea Volcano, Hawaii	Imperial Valley, Calif.	Marysville, Montana

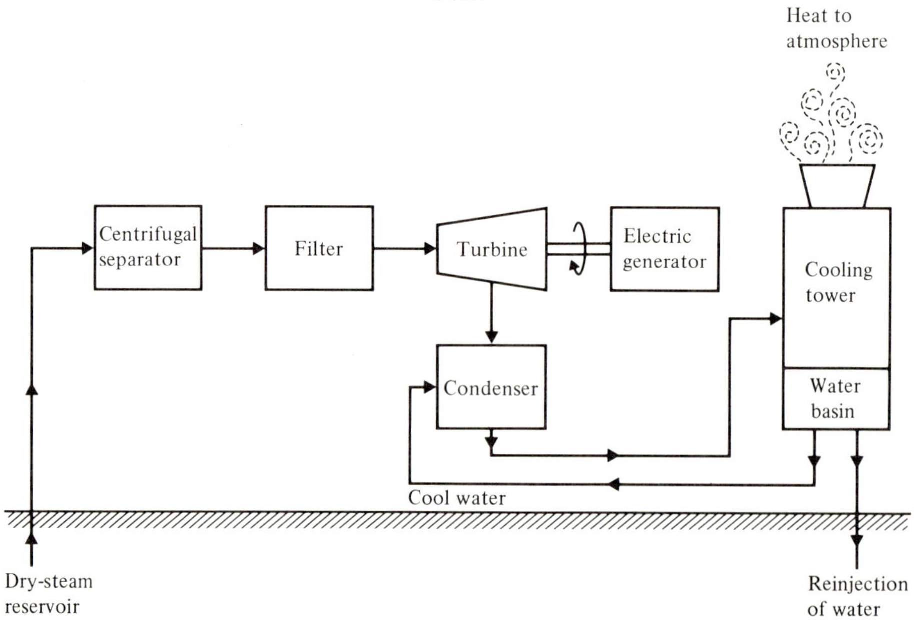

Schematic diagram of a dry steam electric power plant using a cooling tower and a reinjection well. This plant operates with an efficiency of 22%.

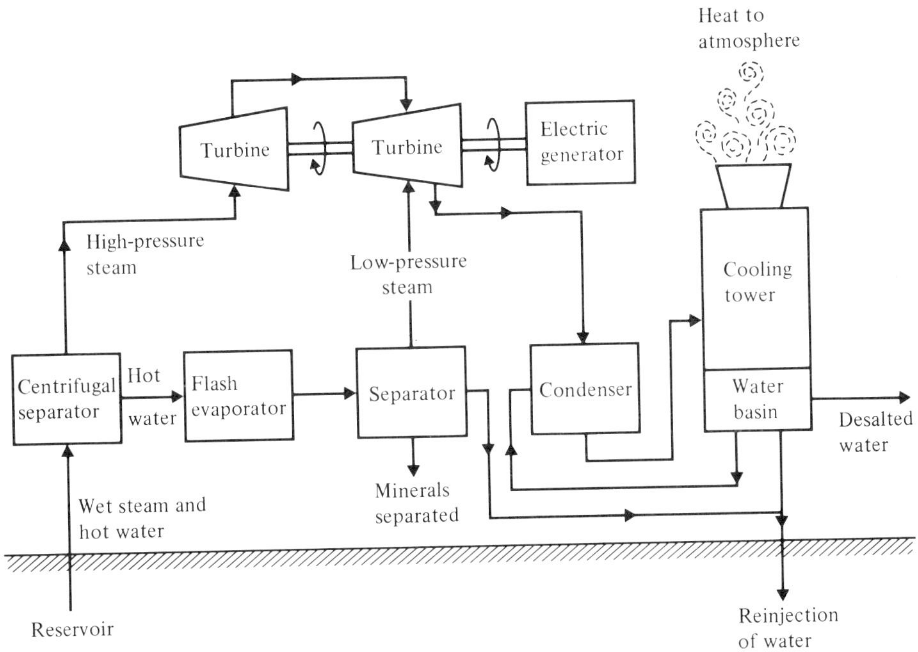

Schematic diagram of a geothermal power plant using wet steam and hot water. Typically the output of the well has a ratio of 1:3 of steam to water by weight.

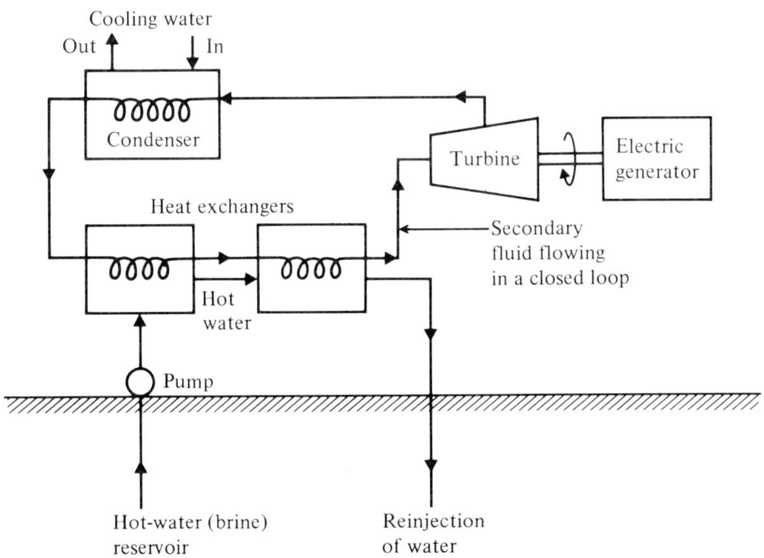

Schematic diagram of a power plant using a separate secondary fluid such as isobutane or freon (a binary cycle system). This system is necessary for hot brine geothermal fields such as that of Imperial Valley, California. The hot brine is too corrosive to be used directly in the turbine and a secondary fluid is utilized.

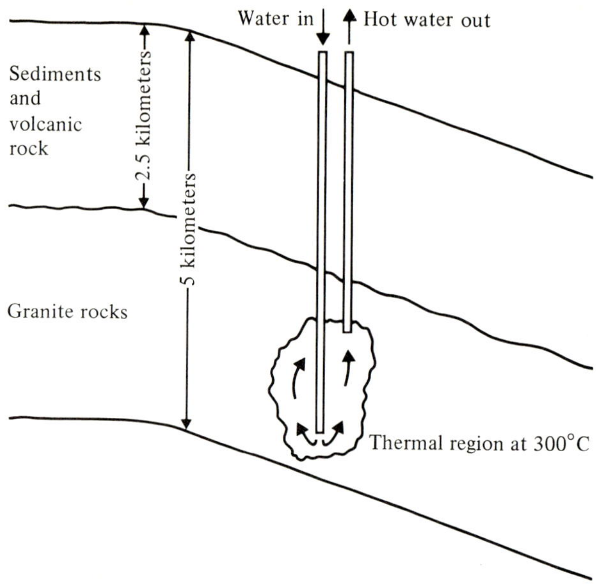

A proposed scheme for extracting energy from a hot rock site. The crack is not a three-dimensional cavity, but rather lies in a vertical plane.

12

NUCLEAR FISSION POWER AND NUCLEAR FUSION RESEARCH

Of all the major new sources of energy, nuclear fission has received the most financial support since 1945 and is correspondingly the most well-developed technology. The development of nuclear fission power dates back to the 1940s with the development of the first atomic explosion at Alamogordo in New Mexico in July, 1945. After World War II, the U.S. formed the Atomic Energy Commission (A.E.C.) in 1946 to oversee the development of nuclear power reactors as well as nuclear weapons. The A.E.C. was split in 1975, and the regulatory portion of the A.E.C. responsibility was assigned to the new Nuclear Regulatory Commission. The other A.E.C. group, the Energy Research and Development Administration (ERDA) was later merged into the Department of Energy. In 1978 the U.S. budget assigned to nuclear fission research and development was one billion dollars.

The recent ERDA energy plan calls for a goal of 6×10^{15} Btu (1.8×10^{12} kWh) in 1985 generated by nuclear fission reactors. However, there is a growing concern in the nation regarding the extent to which research funds are concentrated on nuclear fission reactors and the consequences of large scale use of nuclear fission as a source of electric power.

With nuclear fission reactors slowly growing in number, and under scrutiny regarding their environmental safety, attention is turning to the potential for nuclear fusion. Whereas nuclear fission gets its energy from large atoms as they break up, thermonuclear fusion reactions yield energy when two light nuclei fuse together to form a heavier one. It is fusion that powers the sun and the stars, and the challenge is to establish a workable fusion reactor in practice.

Two light nuclei are brought together to yield energy upon fusion within a high-temperature gas. The very hot gas at a relatively low pressure is known as a *plasma.* Within the plasma are free electrons and ions that flow like a current.

Research and development in the U.S. and Russia on the physics of fusion has increased over the past several years. The U.S. Energy Research and Development Administration provided funds to researchers in the amount of \$75 million and \$450 million in 1973 and 1979 respectively. It is estimated that the USSR and the USA will each spend over 1.5 billion dollars over the period 1975 to 1980 on fusion research. Fusion power research is proceeding with success and it is expected that commercial fusion power would not be available before the 21st century.

THE FISSION PROCESS

The process of fission can be illustrated as process of ^{235}U abosorbing a neutron, splitting into barium and krypton and releasing 3 more neutrons as follows:

$$^{235}U + n \qquad ^{92}Kr + ^{141}Ba + 3n$$

Accounting for the total mass on both sides of the equation, the loss of mass after splitting ^{235}U is 0.215 amu* (3.57×10^{-28} kg). The energy released is 3.2×10^{-11} J for one fission. If 1 percent of 2.5×10^{24} atoms in a kilogram of uranium-235 were to fission over a very short time, then the total energy would be 8×10^{11} J.

*amu = atomic mass units = 1.66×10^{-27} kg.

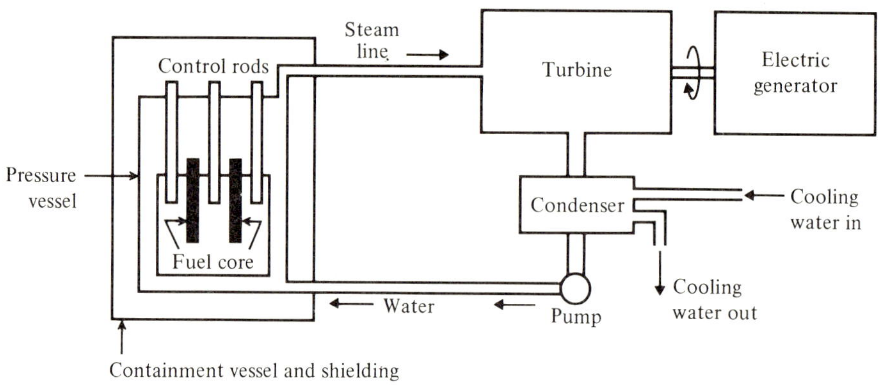

A schematic diagram of a water-steam circuit in a boiling-water reactor. The core consists of bundles of fuel rods which contain uranium oxide pellets. A majority of today's reactors are boiling-water reactors (BWR). In a BWR, water is used as the coolant and serves also the moderator. The water is brought into the reactor and allowed to boil. It then is taken out as pressurized steam and used to drive the turbine. Typically the BWR operates at 7×10^5 kgs/m^2 (1,000 lbs/in^2) pressure and a temperature of 300°C. The BWR has the advantage of simplicity, but suffers from the disadvantage of the potential of radiation passing to the turbine via the water. Also, the efficiency of this system is only about 30%.

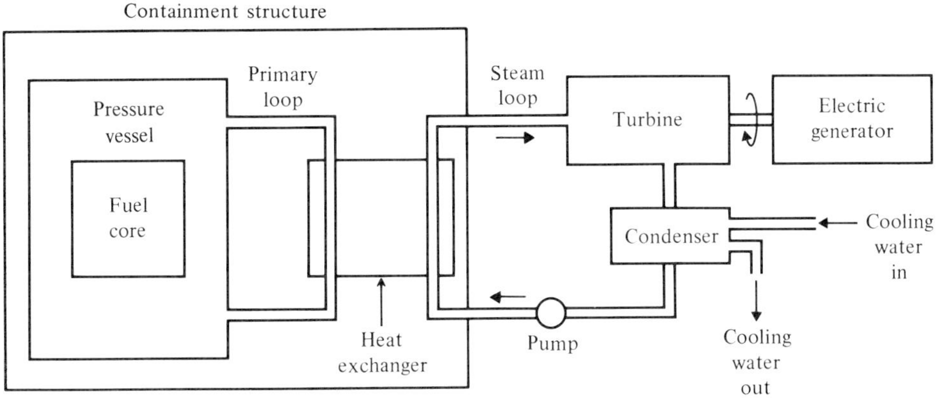

A schematic diagram of two-loop system using a steam generator to provide the steam to drive the turbine. The fluid or gas in the primary loop may be water, helium gas, or carbon dioxide gas. When water is used, the system is called a pressurized water reactor (PWR).

The Yankee Atomic Power Plant at Rowe, Massachusetts. The containment vessel can be seen behind the electric generating building. Photo courtesy of Yankee Atomic Electric Company.

NUCLEAR POWER CAPACITY IN THE U.S.

Year	*Capacity (Gigawatts)*	*Number of Plants*
1957	0.09	1
1960	0.290	2
1965	0.891	7
1970	4.9	14
1972	14	
1973	21	
1975	39.9	58
1978	54.1	72
1980*	85	100
1985*	130	

*Projected

NUCLEAR ENERGY PRODUCTION IN THE U.S.

Year	*Production ($\times 10^{12}$ Btu)*	*(10^6 kw-hrs)*
1957	1	10
1960	6	518
1965	38	3,657
1970	229	21,801
1971	391	37,899
1975	1,839	172,505
1978	3,012	275,000

ANNUAL PRODUCTION OF URANIUM ORE

	Production (Thousand Tons)	
Year	*United States*	*World*
1960	17.8	43
1965	10.3	19
1970	12.9	
1973	13.2	25
1976	17.0	
1980*	30.3	80
1985*	45.0	135

*Projected

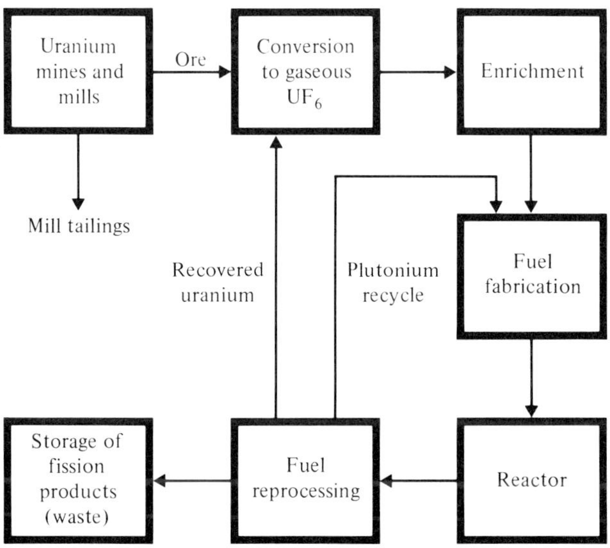

The fuel cycle for reactors using water as a coolant. The enrichment process increases the ^{235}U content from 0.7% to 2 or 3% for ultimate use in a reactor. An enrichment plant costs about one billion dollars to build. A full-fuel load for one new 1000-MW reactor costs about $38 million and replenishment every year or so costs another $11 million or about $400 million over the life of the plant. The question of the ultimate disposal of the radioactive wastes remains unsolved at present.

DOMESTIC RESOURCES OF URANIUM IN THE U.S.

	(Thousands/Tons)		
Cost	*Reasonably Assured*	*Estimated Additional*	*Total*
up to \$20/kg	315	960	1,275
up to \$30/kg	420	1,500	1,950
up to \$60/kg	600	2,900	3,500

Source: M. C. Day, "Nuclear Energy: A Second Round of Questions," *Bulletin of the Atomic Scientists,* Dec. 1975, pp. 52–59.

THE CAPITAL COSTS OF A NUCLEAR FISSION REACTOR

Year of 1st Operation	*Planned or Budgeted Cost (\$/kW)*	*Actual Cost (\$/kW)*
1966	\$ 120	\$120
1968	165	165
1970	150	220
1972	240	400
1974	310	520
1975	420	610
1980	700	
1985	1,100	

NEW ORDERS FOR NUCLEAR REACTORS IN THE U.S.

Year	1969	1973	1974	1976	1977	1978
Number Ordered	4	41	28	8	0	2

U.S. SHARE OF THE WORLD NUCLEAR REACTOR EXPORT MARKET

Year	1974	1975	1976	1977
Total Export Order	18	10	6	7
U.S. Reactors Ordered for Export	10	5	1	0

The Liquid Metal Cooled Fast Breeder Reactor (LMFBR) is able to produce more ^{239}Pu from the fertile material ^{238}U than it would consume. It is called a fast reactor since it contains no moderator material to cause a slowdown of the neutrons.

The LMFBR uses a liquid metal, sodium, as the reactor cooler, An inert gas, argon, is used to blanket the sodium. The fuel used in the LMFBR is a mixture of uranium and plutonium. The primary-loop coolant, sodium, conveys the heat to the heat exchanger where it is passed to the turbine loop.

The principle of breeding or generating new nuclear fuel is based on the fact that two man-made isotopes that are fissionable material can be generated. These two isotopes are ^{233}U and plutonium ^{239}Pu. The latter can be made from ^{238}U and since ^{238}U is 140 times as plentiful as ^{235}U in natural uranium ore, the use of plutonium as a fuel becomes a possibility.

Thus, although ^{238}U is not readily fissionable, it converts to ^{239}Pu which is fissionable. This conversion to plutonium also takes place in conventional light-water reactors but not as at a high a rate as in *breeder reactors.* In a breeder reactor, for each atom of fissionable material consumed, more than one atom of fertile material is converted to fissionable material.

The Phenix breeder reactor is located on the Rhone River near Marcoule, France. (Courtesy of the French Information Office.)

The U.S. domestic and export market for nuclear reactors is rapidly declining and the future of the U.S. nuclear reactor industry is in doubt.

ESTIMATED GENERATING COSTS FOR ELECTRIC POWER FOR FUTURE PLANTS IN 1977 DOLLARS

Plant Type	*Cost (Cents/kWh)*
Nuclear	35
High-Sulfur Coal with Scrubbers	42
Low Sulfur Coal without Scrubbers	35
Oil without Scrubbers	42
Calculated for Northern Illinois	

Source: A. D. Rossin and T. A. Rieck, "Economics of Nuclear Power," *Science,* August 18, 1978, pp. 582–589.

NUCLEAR REACTOR ACCIDENTS

Date	*Location*	*Occurence and Consequences*
1952	Chalk River, Ontario, Canada. A low power reactor.	The core melted down due to operator error, but no injury was experienced.
1957	Windscale, England	A failure in operating procedure led to radiation leakage over a 200 square mile area.
1961	Idaho Falls, Idaho, U.S. Government Test Facility.	An experimental reactor failed killing three technicians.
1966	Fermi Experimental Breeder Reactor, Detroit, Michigan	The core partially melted when a sodium cooling system jammed. The emergency systems controlled the reactor but the plant was closed in 1972.
1970	Commonwealth Edison's Dresden II, Morris, Illinois	A malfunction and operator error led to water escaping the reactor vessel into the containment structure.
1975	TVA Brown Ferry, Decatur, Alabama	Two electricians inadvertently started a fire which knocked out the reactor controls. By improvising, the core was kept covered by water.
1979	Three Mile Island, Harrisburg, Penn.	A valve failed resulting in a chain of mechanical and human failures. The reactor was shut down but radioactive steam leaked and spread over an area of up to 20 miles. The reactor will take years and millions of dollars to clean up and bring back into operation. The radiation released in the immediate area was about 20 millirem/hour for a few hours and lower radiation for up to ten days*. This radiation may result in one or two cancer deaths or injuries.

*The rem is the unit of dose equivalent of radiation.

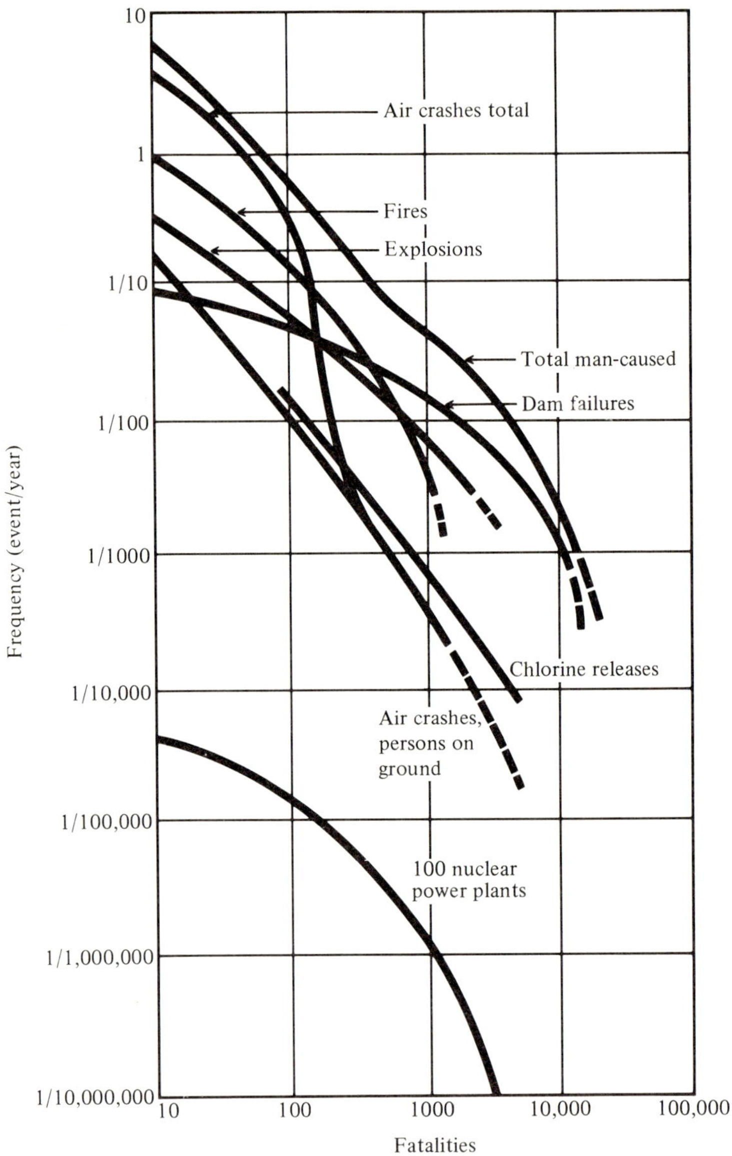

Frequency of fatalities due to man-caused events. *Source:* Reactor Safety Study, U.S. Nuclear Regulatory Commission, WASH-1400, October 1975, Washington, D.C.

The Reactor Safety Study, known as WASH-1400, involved 70 man-years of effort and some $4 million, and was published in 1975. Several critiques of the study have faulted the calculations and the study process. The Nuclear Regulatory Commission stated:

"Despite its shortcomings, WASH-1400 provides at this time the most complete single picture of accident probabilities associated with nuclear reactors. The fault-

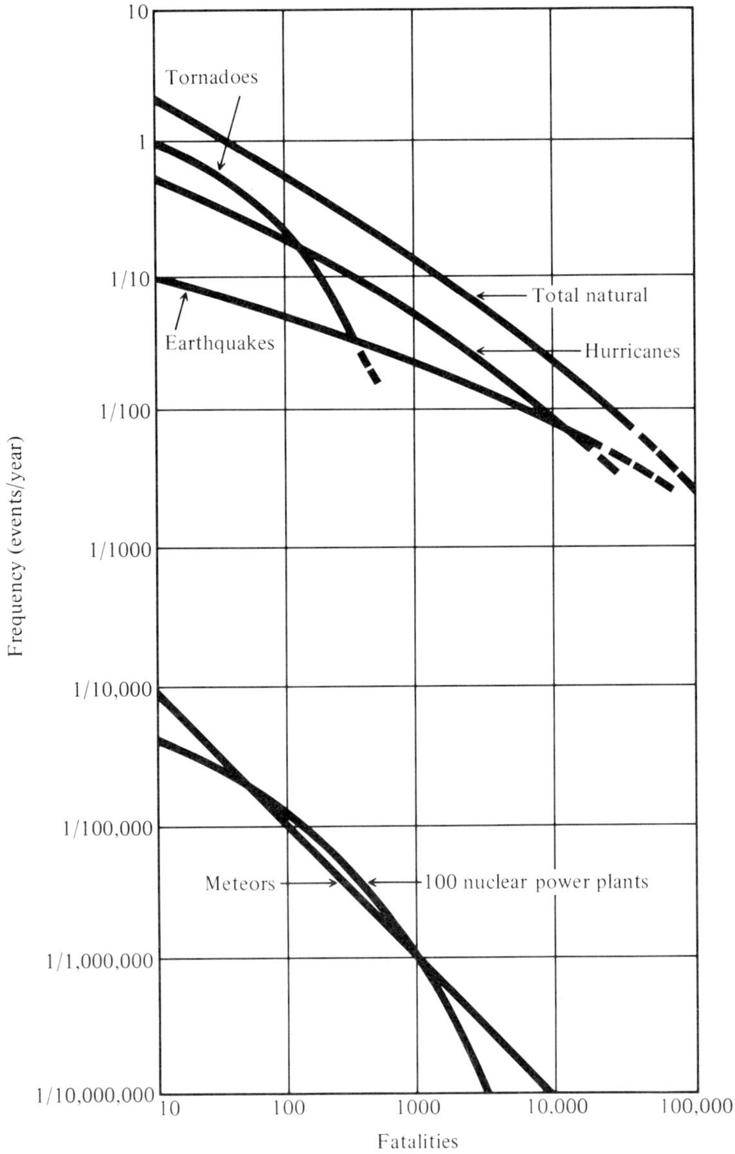

Frequency of fatalities due to natural events. *Source:* Reactor Safety Study, U.S. Nuclear Regulatory Commission, WASH-1400, October 1975, Washington, D.C.

tree/event-tree approach coupled with an adequate data base is the best available tool with which to quantify these probabilities. WASH-1400 made clear the importance to reactor safety discussions of accident consequences other than early fatalities."

Nevertheless, the N.R.C. has withdrawn endorsement of the executive summary of the report and acknowledged the limitations of the full report.

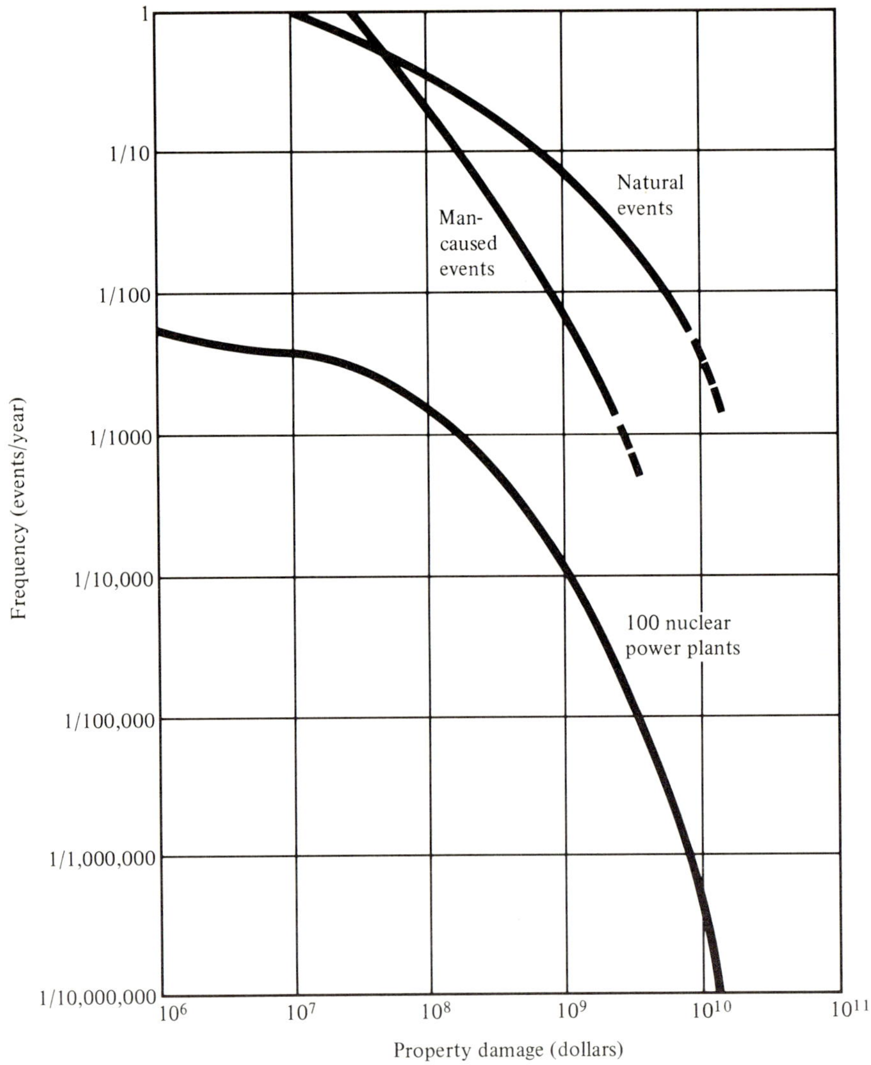

Frequency of property damage due to natural and man-caused events. *Source:* Reactor Safety Study, U.S. Nuclear Regulatory Commission, WASH-1400, October 1975, Washington, D.C.

AVERAGE RISK OF FATALITY BY VARIOUS CAUSES FOR 1969 IN THE U.S.

Accident Type	*Total Number*	*Individual Chance per Year*
Motor Vehicle	55,791	1 in 4,000
Falls	17,827	1 in 10,000
Fires and Hot Substances	7,451	1 in 25,000
Drowning	6,181	1 in 30,000
Firearms	2,309	1 in 100,000
Air Travel	1,778	1 in 100,000
Falling Objects	1,271	1 in 160,000
Electrocution	1,148	1 in 160,000
Lightning	160	1 in 2,000,000
Tornadoes	91	1 in 2,500,000
Hurricanes	93	1 in 2,500,000
All Accidents	111,992	1 in 1,600
Nuclear Reactor Accidents (100 Plants)	—	1 in 5,000,000,000

Source: Reactor Safety Study, U.S. Nuclear Regulatory Commission, WASH-1400, October 1975, Washington, D.C.

THE WORST CASE NUCLEAR ACCIDENT

Probability = 10^{-9} per Reactor Year
Early Cancer Fatalities = 3,300
Radiation Illness = 45,000
Property Damage = \$14 Billion
Delayed Cancer Fatalities = 1,500 per Year for Up to 10 Years

Source: Reactor Safety Study, U.S. Nuclear Regulatory Commission, WASH-1400, October 1975, Washington, D.C.

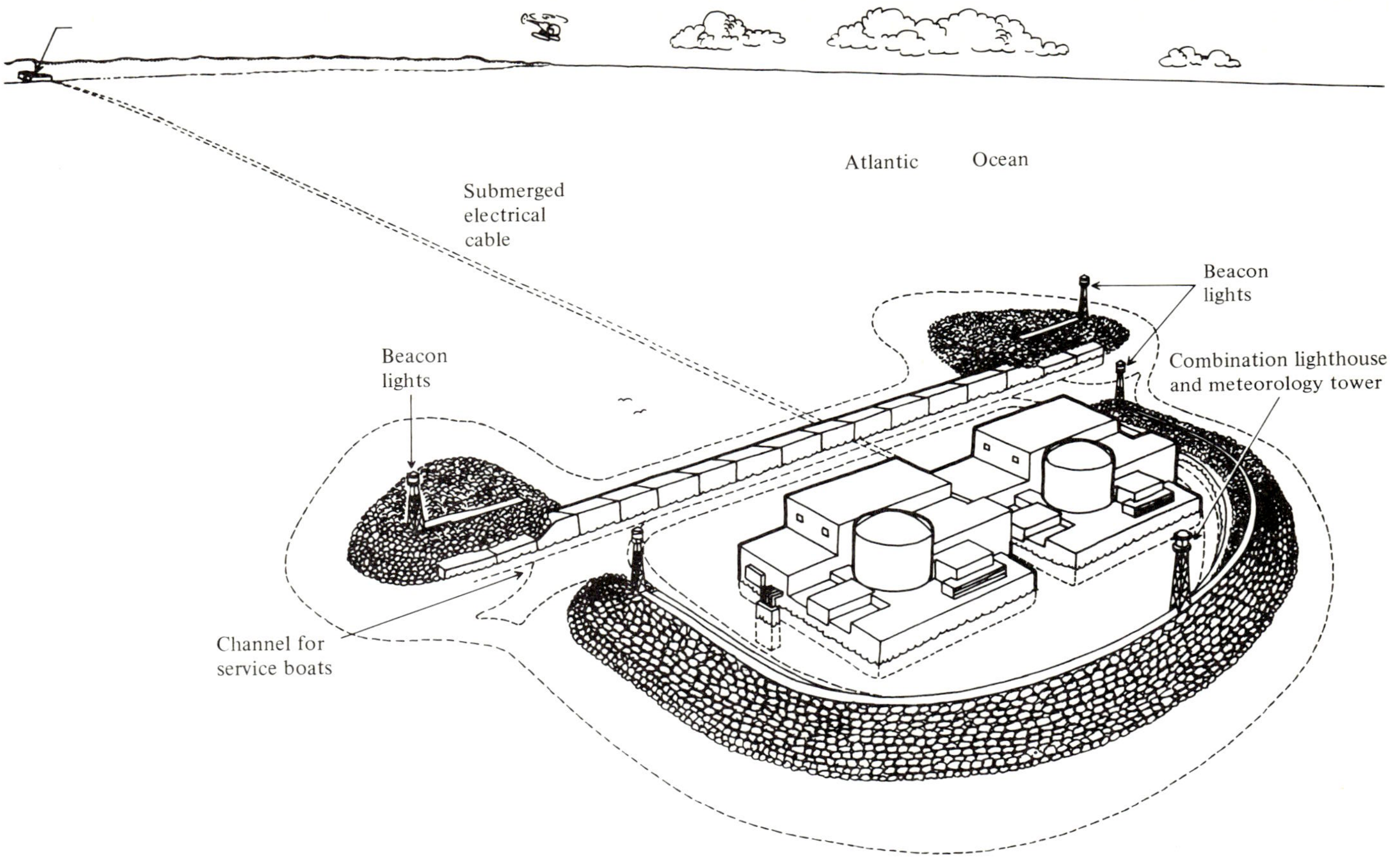

A drawing of the floating nuclear plant concept. In order to improve the safety and reduce the environmental effects of nuclear power plants it may be wise to place the plants underground or off the coast.

NUCLEAR POWER IN THE WORLD

Nation	1974 Capacity (Gigawatts)	1974 Number of Plants	1980 Capacity (Gigawatts)	1980 Number of Plants
U.S.	38.0	58	85	100
USSR	2.6	9	15	25
Japan	3.1	7	32	40
Canada	2.5	6	7.5	13
Great Britain	5.6	14	12	19
France	2.8	11	15.0	23
W. Germany	2.3	11	20	30
WORLD TOTAL	57.0	134	200	450

Note: A plant may contain one or more reactors.

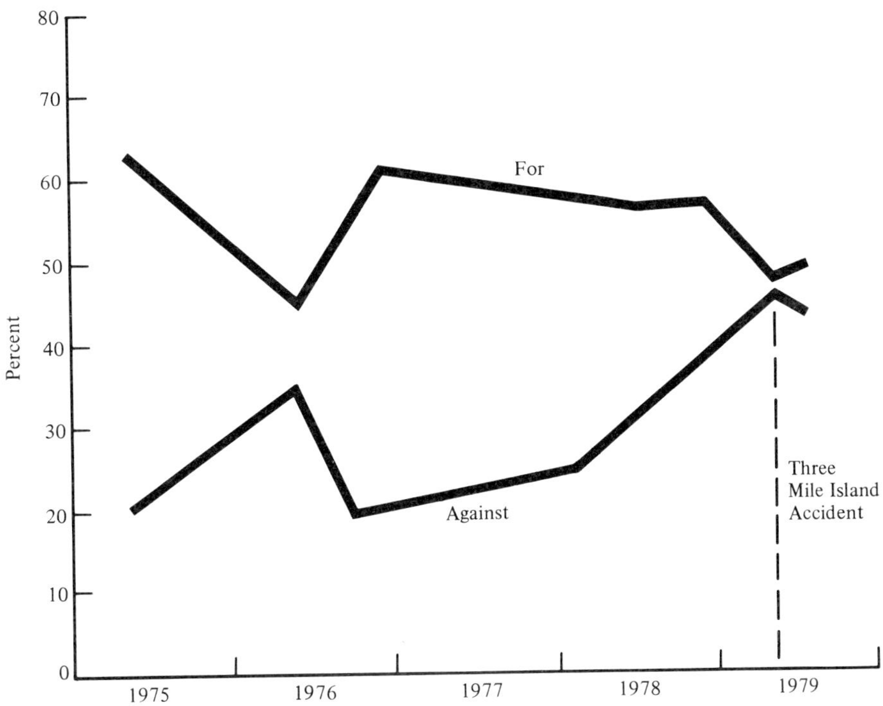

The ABC News-Harris Survey asked the U.S. population the question "Do you favor or oppose the building of new nuclear power plants in the U.S.?" In March 1975, 63% favored and 19% opposed new plants. However, after the Three-Mile Island plant accident, the ratio dropped to 47% to 45%. (*Source: Fortune,* Sept. 10, 1979, pg. 128.)

THE FUSION REACTION – THE DEUTERIUM – TRITIUM REACTION

Lithium is used to reproduce or breed the tritium fuel.

$$D + T \qquad He^4 + n + 17.6 \text{ Mev}$$

$$n + Li^6 \qquad He^4 + T + 4.8 \text{ Mev}$$

The neutron, n, combines with lithium in the second equation to yield helium and tritium plus released energy. The tritium is used again in the process of the equation, thus recycling itself. Combining the equations we have the net reaction as

$$D + Li^6 \qquad 2He^4 + 22.4 \text{ Mev.}$$

The D-T reactor requires a second reaction to take place outside the plasma, and this is technologically difficult. In the sequence of the fusion process, the electrons are stripped from the atoms, leaving them positively charged. Particles with like charges repel each other. At very high temperatures, the particles become energetic enough to overcome their repulsion, collide, and fuse. About 80% of the resultant energy release carries off in a 14.1 MeV neutron, the balance in the helium-4 nucleus and x-ray. In order to achieve a net positive release of fusion energy, the product of particle density, n, (in particles per cubic centimeter) and confinement time, τ, (in seconds) must exceed 10^{14}. This is called the Lawson criterion and is written as

$$n\tau > 10^{14} \text{ atoms/sec/cm}^3.$$

The twin requirements of high kinetic temperature and adequate confinement times define the challenge to fusion research. Reasonable goals for a steady-state reactor are: plasma density $n = 5 \times 10^{14}$ atoms/cm^3; pulse duration $\tau = .5$ second; and plasma temperature T = 10 kev.

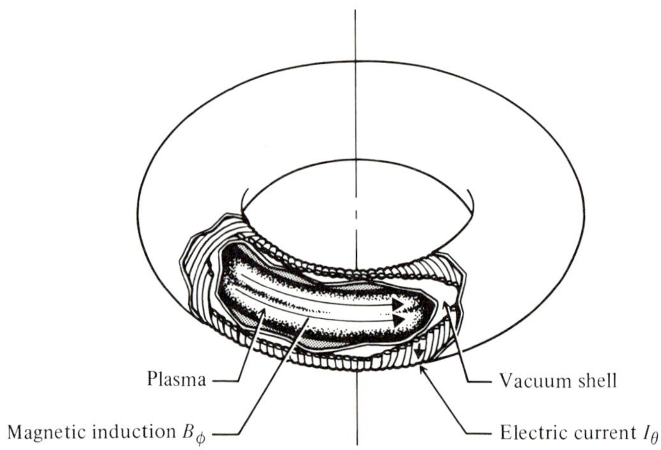

A torodial magnetic field B_θ made by the current I_θ. This scheme is used for magnetic confinement of the plasma of nuclei and electrons to the doughnut-shaped chamber.

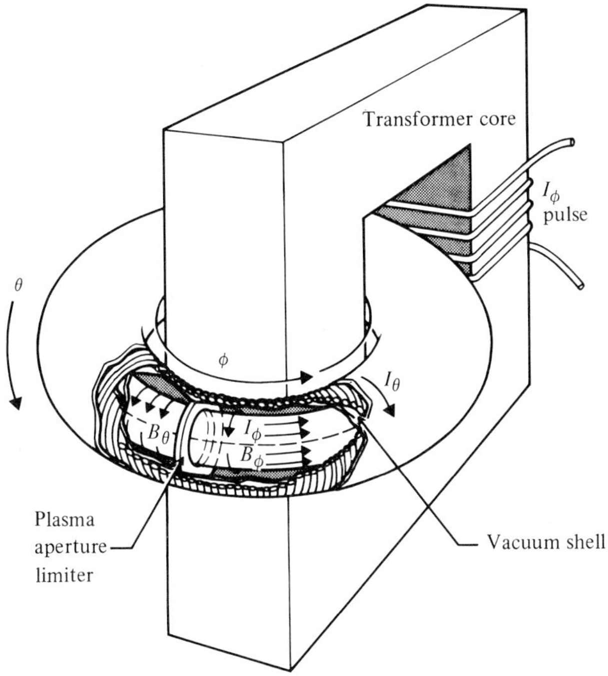

The Tokomak plasma-confinement scheme. Tokomak systems are operating at the Kurchatov Institute in Moscow and Princeton University in New Jersey.

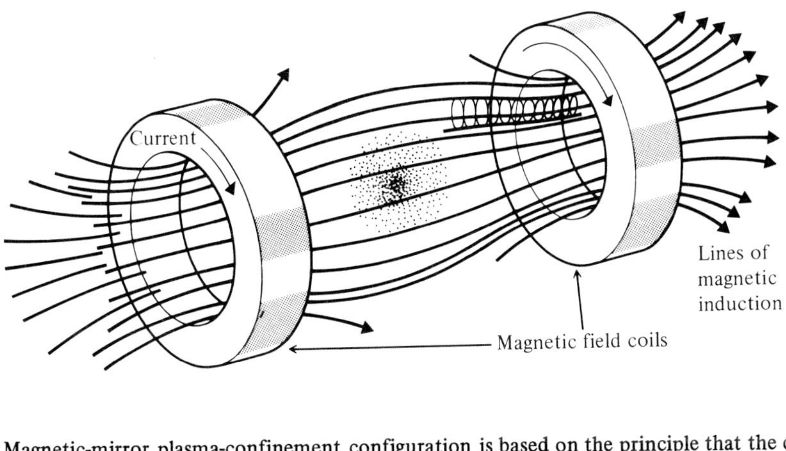

Magnetic-mirror plasma-confinement configuration is based on the principle that the charged particles in a plasma will be reflected at regions of very strong magnetic field at the two ends of the confinement volume.

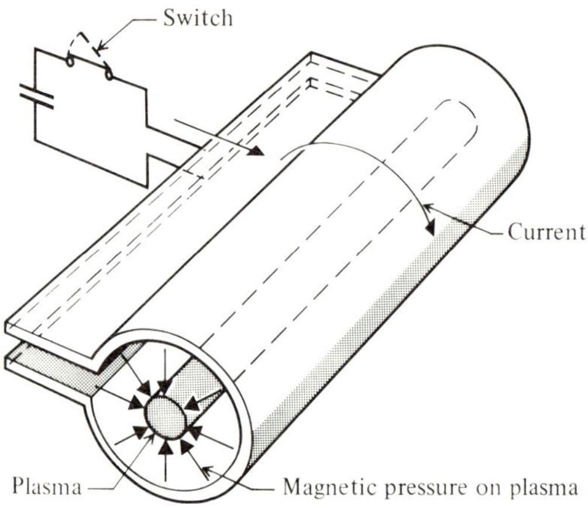

Theta-pinch plasma-confinement scheme. The confining field in the θ axis holds the plasma within the chamber.

PERFORMANCE OF CONFINEMENT SCHEMES

	Tokomak	*Mirror*	*Theta Pinch (Scylla)*	*Lawson Criterion*
Plasma Density (ions/cm^3)	10^{14}	10^{14}	10^{17}	5×10^{14}
Plasma Temperature (°C)	4×10^7	10^8	10^7	10^8
Confinement Time (msec)	50	1	0.01	500
Problems	Heating	Reducing Losses Through Mirrors	Stability	

Source: J. P. Holdren, "Fusion Energy in Context," *Science,* April 14, 1978, pp. 168–180.

The Lawson criterion states that the reactor can produce net power when the confinement parameter $n\tau$ exceeds 10^{14} cm^{-3} sec at an ion temperature around 10^8 °C (n is the fuel ion number per cubic centimeter and τ is the mean confinement time in seconds). The basic idea embodied in the Lawson criterion applies to pinch confinement as well as to magnetic confinement, but where the latter might achieve the break-even condition with $n = 10^{14}$ cm^{-3} and $\tau = 1$ second, the former would do so with perhaps $n = 10^{26}$ cm^{-3} and $\tau = 10^{-12}$ second.

The Lawrence Livermore Laboratory is pursuing experiments using the magnetic mirror approach. Plasma guns fire a burst of preheated fuel into the device where the magnets contain and squeeze the fuel. A neutral beam of 250 million degrees F is then injected. (a) The schematic diagram of 2XIIB. (b) The actual experiment. At lower left are the plasma guns. The domed tanks house the fusion fuel injection system.

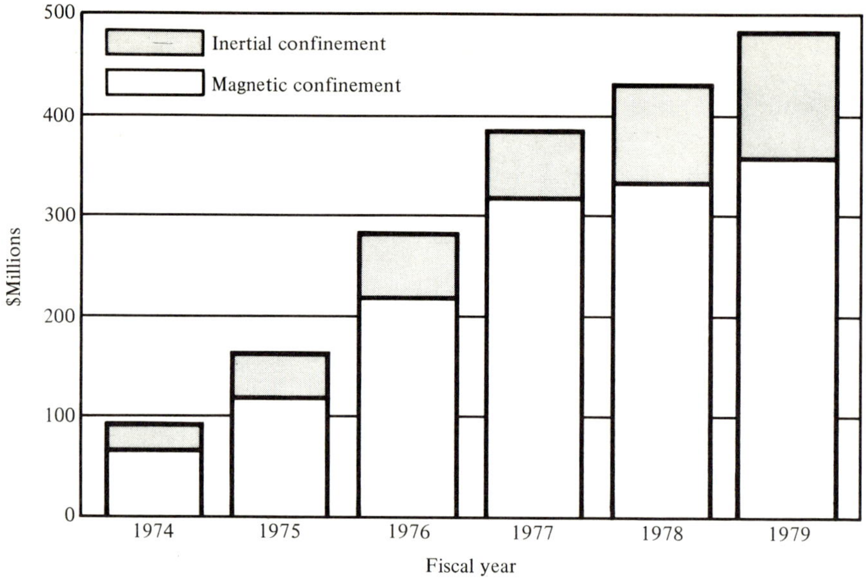

Federal fusion R&D budget over the period 1974–1979. *Source:* U.S. Dept. of Energy.

13

THE WIND AND THE OCEANS

A large amount of energy is stored in the wind and the oceans. The Persians built the first-known windmills as early as 250 B.C. and several are still in use today. Their sails revolve around a vertical axis like carousels. Windmills were continually refined over centuries, but fell to disuse with the introduction of fossil-fuel fired machines and electric power. Now, we are wisely reexamining the potential available in the energy of the wind.

Energy is stored in the tides, currents, and thermal grandient of the oceans. The tides and tidal currents have a power estimated at 3×10^{12} watts. The energy stored in the thermal layers of the ocean as energy received from the sun is vast, but widely dispersed over the surface of all the oceans of the world. There is also energy present in the wave motion of the surface of the oceans. Tapping the energy of the ocean is difficult and costly due to the necessity of collecting the widely dispersed energy. The best opportunity for using the energy of the oceans lies, at present, in utilizing the tidal energy sources of the world.

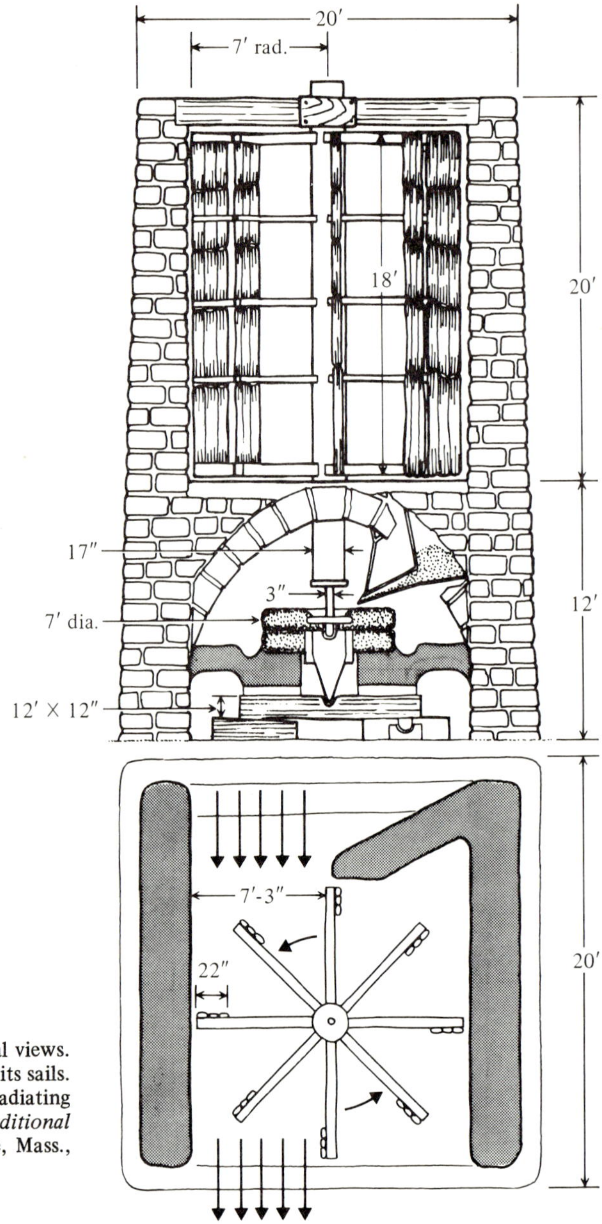

Persian vertical windmill with grindstone, in sectional views. This version has bundles of reeds instead of cloth for its sails. The wind is caught by sails rigged between spokes radiating from the top and bottom of the shaft. [*Source: Traditional Crafts of Persia,* Hans Wulff (MIT Press, Cambridge, Mass., 1966)]

A sketch of a horizontal Dutch windmill. The blades rotate around a horizontal axis with the plane of the planes essentially perpendicular to the wind velocity sector. The horizontal windmill, whose sails rotate around a horizontal axis, spread throughout the Islamic world after the Arab conquest of Iran. The windmill in Europe in the 11th century was also a horizontal axis type and by the 17th century the Netherlands had become the world's most industralized nation by extensive use of wind power in ships and the familiar Dutch windmill.

The Smith-Putnam Wind Machine on Grandpa's Knob in Vermont was built by Palmer Putnam. The two-blade propeller was 175 feet from tip to tip. Each blade weighed 8 tons and consisted of stainless steel. The machine was rated to produce 1,250 kilowatts. The machine was actuated on August 29, 1941, and operated usefully for two years. A bearing failed in 1943 and, since World War II was in progress and this project had low priority, it took several years to obtain a new main bearing. The new bearing was installed early in 1945. Following the installation, the machine was operated only a few months when one blade failed due to overstress after being idle for two years. The machine was subsequently dismantled. The project had cost 1.25 million dollars. The concept was that, if several similar machines could be produced by Smith-Putnam that the cost could be reduced to $200 per kilowatt for construction.

WIND POWER ELECTRIC GENERATORS IN THE WORLD

Name and Location	*Electric Power Rating (kW)*	*Swept Circle Diameter (Feet)*
Smith-Putnam (1941) Vermont, USA	1,250	175
Gedser (1957) Denmark	200	79
NASA-Lewis (1975) Ohio, USA	100	124
NASA (1978) Clayton, New Mexico	200	125
Tvind (1978) Denmark	2,000	178
NASA (1979) Boone, North Carolina	2,000	200

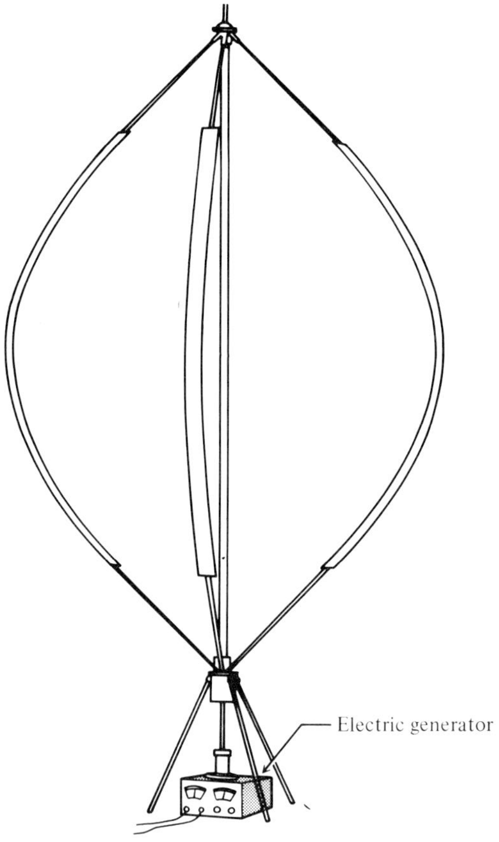

The Darrieus vertical-axis device. The shape of the blades is the natural shape a flexible cable would assume if spun about a fixed vertical axis. The device, patented by G. J. M. Darrieus in 1927, consists of two thin airfoils with one end mounted on the top end of the vertical shaft and the other at the bottom end of the shaft. The resulting configuration is nearly circular in shape and looks similar to an eggbeater. This device is currently under study for use to supply power to single family dwellings.

USABLE ENERGY POTENTIALS ON A GLOBAL BASIS

	(10^{12} *Watts*)	10^{15} *Btu/Year*
Wind	1.3	39
Hydropower	2.9	86
Geothermal	0.13	4
Tidal	0.06	1.9

Wind figure assumes feasible utilization of 1% of global wind resource.

Source: M. R. Gustavson, "Limits to Wind Power Utilization", *Science,* April 6, 1979, pp. 13–17.

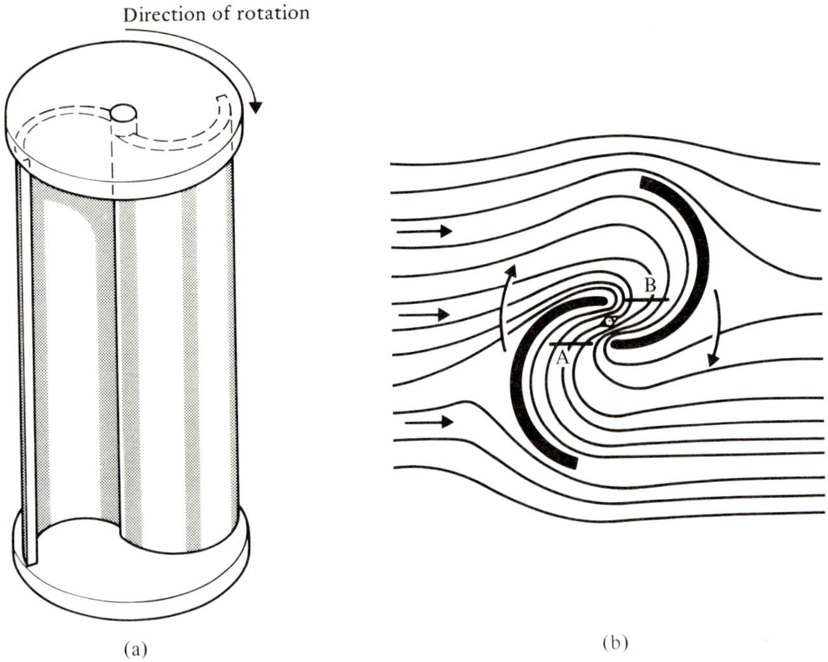

The Savonius rotor. (a) The two semicylindrical vanes joined together on a vertical shaft. (b) A top view showing the flow of air and the rotation of the device around a vertical shaft. The Savonius rotor was formed by cutting a cylinder into two semicylindrical surfaces, moving these surfaces sideways to form an S shaped rotor, and then placing a vertical shaft in the center. End plates are used to provide structural strength. The Savonius rotor (S rotor) operates as a two stage turbine whenever the wind impinging on the concave side is circulated to the center of the rotor to the back of the convex side. The S rotor is particularly adaptable to remote areas and areas where the wind direction is variable.

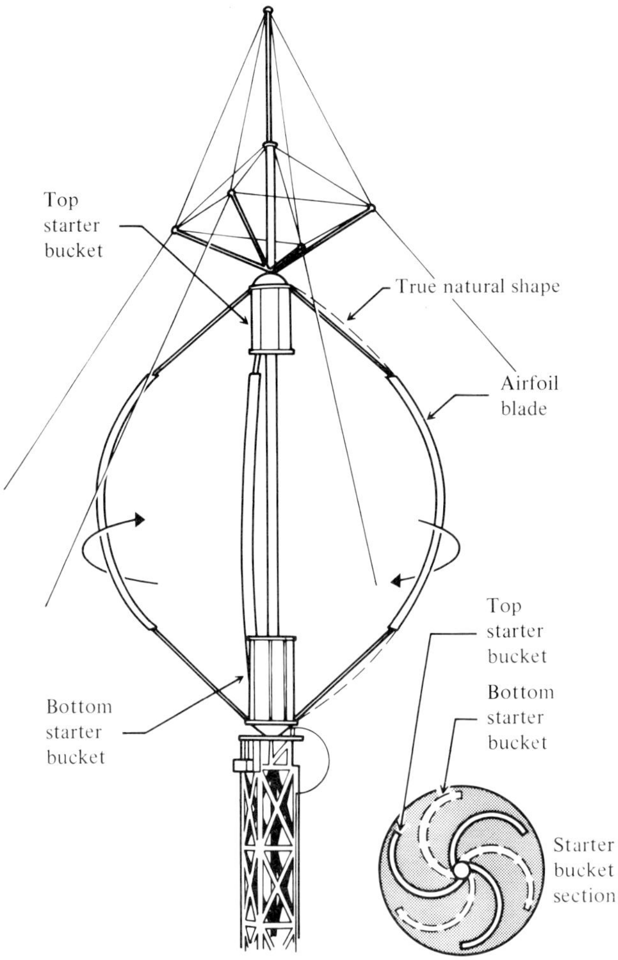

Researchers at Sandia Laboratory in New Mexico are exploring a combination Savonius and Darrieus machine. Cables stabilize the tower in the wind. The airfoil blade is fabricated from three straight pieces and varies from the natural airfoil shape slightly as shown. Small S-rotor devices are installed at the top and bottom of the shaft to provide a self-starting feature for the device.

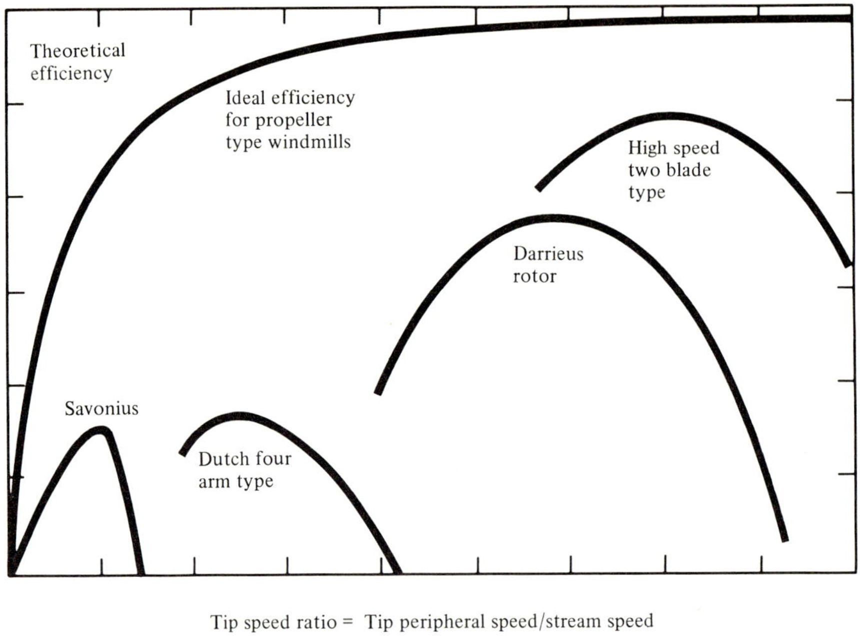

The theoretical efficiency, η, of several types of windmills where $\eta = P_{out}/P_{in}$. The actual efficiencies are about one-half of that shown.

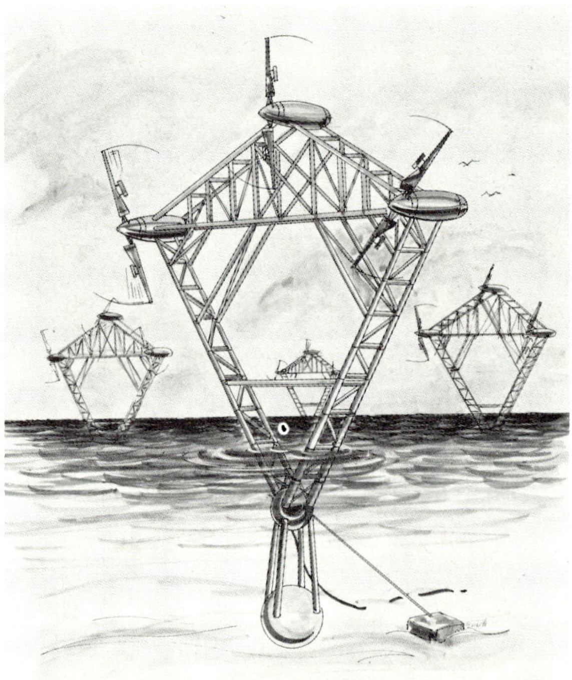

Professor William Heronemus has proposed a system of offshore wind-power machines. The offshore system places three rows of wind stations across the prevailing westerlies in the Gulf of Maine–Georges Bank area of the continental shelf. Each wind machine has a capacity of 2 megawatts, and 3 wind machines are mounted on each station mounted on floating platforms near the 100-meter depth contour. Wind stations are clustered around an electrolyzer plant to yield hydrogen stored in a collector. There are 2,000 such wind platforms totaling to 12×10^6 kilowatts. The average wind speed at towers of 100 feet would be 20 miles/hour. A 200-foot diameter wind machine with three sets of such units mounted on each platform is proposed. The total project is estimated to cost 22 billion dollars. (Courtesy of W. Heronemus.)

A mountaintop overlooking the city of Boone, North Carolina, is the site for the 2,000-kilowatt windmill built by General Electric Company for the Department of Energy and NASA. At an ideal wind speed of 25 mph, the turbine's two 100-foot blades produce 2,000 kW. The 5.8 million dollar machine was dedicated on July 11, 1979. The turbine on top of a 140-foot tower operates for wind speeds in the range of 11 mph to 35 mph.

CHARACTERISTICS OF TIDAL POWER SITES

Site	*Tidal Range, R (Meters)*	*Basin Area A (km^2)*	*P_{actual} (Megawatts)*	*Barrage Length, L (Meters)*	*L/A (m/km^2)*	*L/P_{actual} (m/megwts.)*
Rance, France	11.4	22.0	160	725	33	4.5
Brest, France	6.4	92.0	211	3,640	40	17.3
Chausey, France	12.4	610.0	5,252	23,500	39	4.5
Savern, Britain	11.5	50.0	370	3,500	70	9.5
Passamaquoddy, USA	7.5	120.0	378	4,270	36	11.3
Bay of Fundy, Cobequid, Canada	11	700	4,743	8,000	11.4	1.7
Kislogrub, USSR	4	2	1.8	30.5	15	16.9

Tidal range R = difference in amplitude between high and low tides.
P_{actual} = actual power output
Barrage = the dam across the opening of the bay.

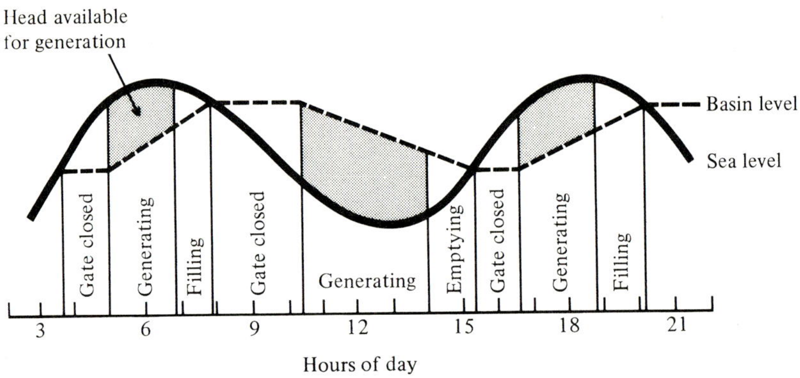

Tidal power is obtained from the oscillatory flow of water in the filling and emptying of partially-enclosed coastal basins during the semidiurnal rise and fall of the oceanic tides. This energy may be partially converted into tidal-electric power by enclosing such basins with dams to create a difference in water level between the ocean and the basin, and then using the water flow while the basin is filling or emptying to drive hydraulic turbines propelling electric generators. Shown in the figure is the cycle of the tides and the generation of electrical power using a two-way flow scheme. The dotted line is the level of the water in the basin and the solid line is the level of the sea.

The Rance River project in France was completed in 1966 for a cost of $100 million. The dam is 725 meters long. A sluice section is incorporated in the dam structure to improve power station operation by speeding up the emptying and filling of the reservoir during the operating cycles. The gates are operated several times a day, and since they are subjected to head reversals, have both upstream- and downstream-sealing arrangements. Both incoming and out-going tides are utilized by using a reversible axial flow built turbine. This turbine is a horizontal Kaplan-type turbine with adjustable wicket gates and runner blades. The project has 24 10,000 kw turbine-generator sets for a total installed capacity of 240 megawatts. The average actual power generated is estimated to be 160 megawatts. In the figure: (a) the power plant is shown under construction; (b) the tidal dam with a road running along the top of the dam. The power plant appears on the left side of the dam. (Photos courtesy of the French Embassy.)

(a)

(b)

14

ENERGY FROM WASTES AND BIOMASS

All human and industrial processes produce wastes that are the normally unused and undesirable products of a specific process. The waste products of a home include paper, tin cans, aluminum cans, and food scraps as well as sewage. The waste products of industry and commerce include paper and wood, metal scraps, as well as agricultural waste products. The U.S. produces approximately 0.9 tonne (one tonne = 1,000 kilograms) of solid waste per person per year or over 185 million tonnes in total. This quantity has been growing at an annual rate of 5% per year, and most of this waste is relegated to the land fills and garbage pits near our metropolitan areas.

Waste products of our industrialized society can be used to provide fuels for our electric power and steam plants. Also, valuable materials now classified as wastes can be recycled. The recycling of metals, such as scrap aluminum or steel, can be energy-conserving since more energy is consumed in the original manufacture of these metals than is consumed in the recycling process.

Another source of fuel is organic biological materials such as agricultural crops, grown specifically to serve as a feedstock, to be converted to methane, hydrogen, electricity or other energy carriers.

AMOUNTS OF AVAILABLE DRY ORGANIC WASTES IN THE U.S. IN 1971 IN MILLION TONS (2 × 10^9 lbs.)

Source	*Wastes Generated*	*Readily Collectable*
Manure	200	26.0
Urban Refuse	129	71.0
Logging and Wood Manufacturing Residues	55	5.0
Agricultural Crops and Food Wastes	390	22.6
Industrial Wastes	44	5.2
Municipal Sewage Solids	12	1.5
Miscellaneous	50	5.0
TOTAL	880	136.3
Net Oil Potential (10^6 Barrels)	1,098	170
Net Methane Potential (10^9 Cubic Feet)	8.8	1.36

One ton is 2,000 lbs., which is equivalent to 907 kilograms or .907 tonne.

THE PHYSICAL COMPOSITION OF TYPICAL MUNICIPAL WASTE

Category	*Percent*
Paper	50.7
Food Waste	19.1
Yard Waste	–
Metal	10.0
Glass	9.7
Wood	2.9
Textiles	2.6
Leather, Rubber and Plastics	3.3
Miscellaneous	1.7
	100%

The noncombustible materials must be separated out and then the combustible materials can be burned in a boiler for electric generating or steam generating plants. An incinerator plant in Nashville, Tennessee, burns over half of the city's daily refuse of 700 tons and provides heat for the 32 buildings surrounding the facility.

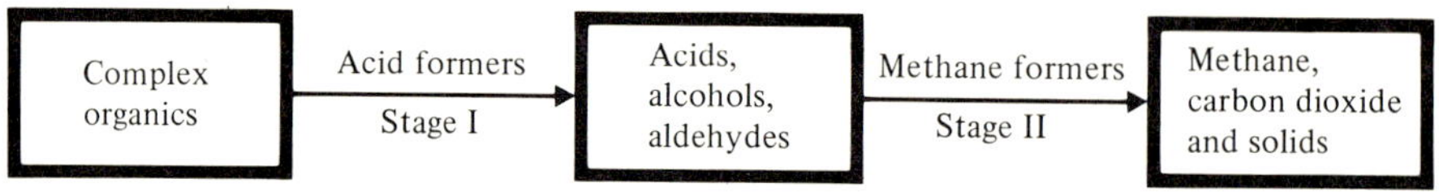

Anarobic digestion occurs in two stages. Acid-forming bacteria are active in the first stage, and methane-forming bacteria in the second stage.

METHODS FOR CONVERTING ORGANIC WASTES OR BIOMASS TO SYNTHETIC FUELS

Method	*Process*	*Product*	*Characteristics*
Hydrogenation	Conversion of Cellulose	Heavy Oil with an Oxygen Content of 10%	Relatively Expensive
Pyrolysis	Thermal Decomposition in the Absence of Oxygen	Oil and Lower Energy Gas	Reasonably Priced, Plant Operating in Baltimore
Anaerobic Digestion	Digestion of Wastes	Methane Gas	Similar to Sewage Treatment Plant

PLANT PRODUCTIVITY

Location	*Plant*	*Annual Production (Metric Tons/Acre)*
California	Algae	20 - 30
Germany	Reedswamp	18.6
Mississippi	Water Hyacinth	4.5 - 13.4
England	Coniferous Forest	13.8
West Indies	Tropical Fruit	23.9
Minnesota	Maize	9.7
Hawaii	Sugar Cane	30.4
Georgia	Sycamore	11 - 16

One acre = 0.405 hectare

ALTERNATIVE FUELS FROM BIOMASS

Fuel	*Product*	*Advantages*	*Disadvantages*
Methanol (Methyl Alcohol)	CH_3OH	Cost similar to gasoline. Can be used in 10% blend with gasoline for autos.	Distribution, risk of separation in the blend
Ethanol (Ethyl Alcohol)	C_2H_5OH	Can be used in gasoline blends up to 10% ethanol. Ethanol has 80,000 Btu/gallon.	Two times as expensive as gasoline (1979).
Biogas	Low Btu Gas, Methane	Clean Burning	Need to modify engines to use.

Biomass is any organic agricultural product. Gasohol is a mixture of 10% ethanol and 90% gasoline.

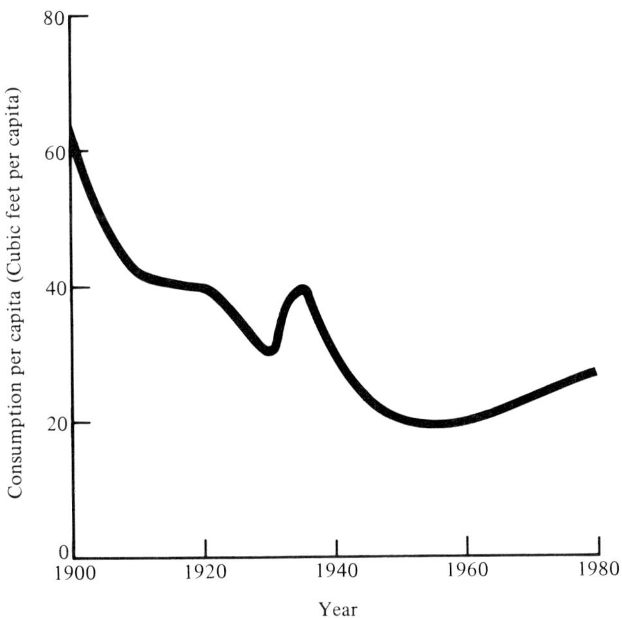

Wood was an important fuel in the U.S. until 1940. Wood may again supply up to 5 or 10 percent of our fuel needs by 2000. The per capita consumption of wood as a fuel is shown for the period 1900 to 1980.

15 SOLAR ENERGY

All our food and fuel (exclusive of nuclear energy) has been made possible by the sun through the photosynthetic combination of water and atmospheric carbon dioxide in growing plants. Solar energy is the basic energy support for life and underlies the wind, the climate, and fossil fuels.

Outside the earth's atmosphere the sun provides energy at the rate of 1,353 watts/m^2 normal to the sun. Since the area of the diametral plane of the earth is 1.27×10^{14} m^2, the solar input to the earth is 1.73×10^{17} watts.

The solar radiation received at the earth's surface is only a small percentage of the total outside the earth's atmosphere. A considerable portion is reflected back into outer space from the atmosphere and clouds. Some solar energy is scattered by water droplets in clouds and dust particles. In cloudy weather the radiation is reduced even further. Solar energy predominantly lies in the spectrum of light waves between 0.3 and 1.2 micrometers.

The average solar radiation received at the earth's surface is approximately 630 watts/m^2. This is an average figure and the actual radiation received at a specific location can be significantly higher or lower. Of course, this radiation is only received at a time the point on earth is normal to the sun's rays. Thus, no radiation will be received at night, and a reduced amount is received at times other than noon. The solar energy received also varies significantly with latitude, as well as with time of year. The solar energy received in a given location in the U.S. during a clear day in December is about one-half of that received in July.

The solar radiation at the earth's surface averaged over the day and over the total year has been recorded for the U.S. and is called the *average annual insolation.*

AVERAGE ANNUAL INSOLATION IN THE U.S.

Location	*Insolation (Watts/m^2)*
Vermont	150
New York City	180
Washington, D.C.	200
Denver	220
San Francisco	230
Arizona – New Mexico	260

A solar-energy-operated printing press at the Paris Exposition of 1878. (Courtesy of National Acronautics and Space Administration.)

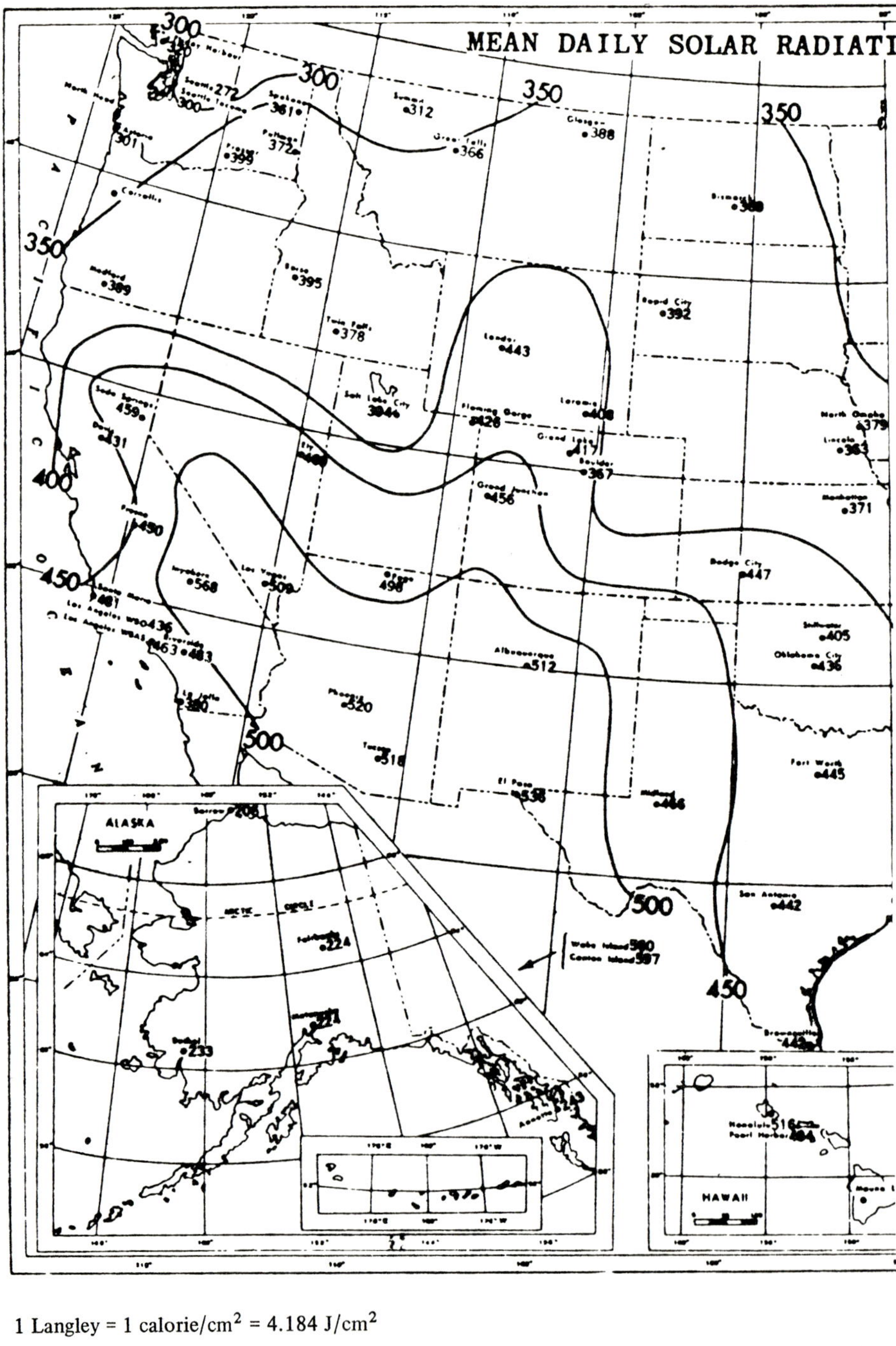

1 Langley = 1 calorie/cm^2 = 4.184 J/cm^2

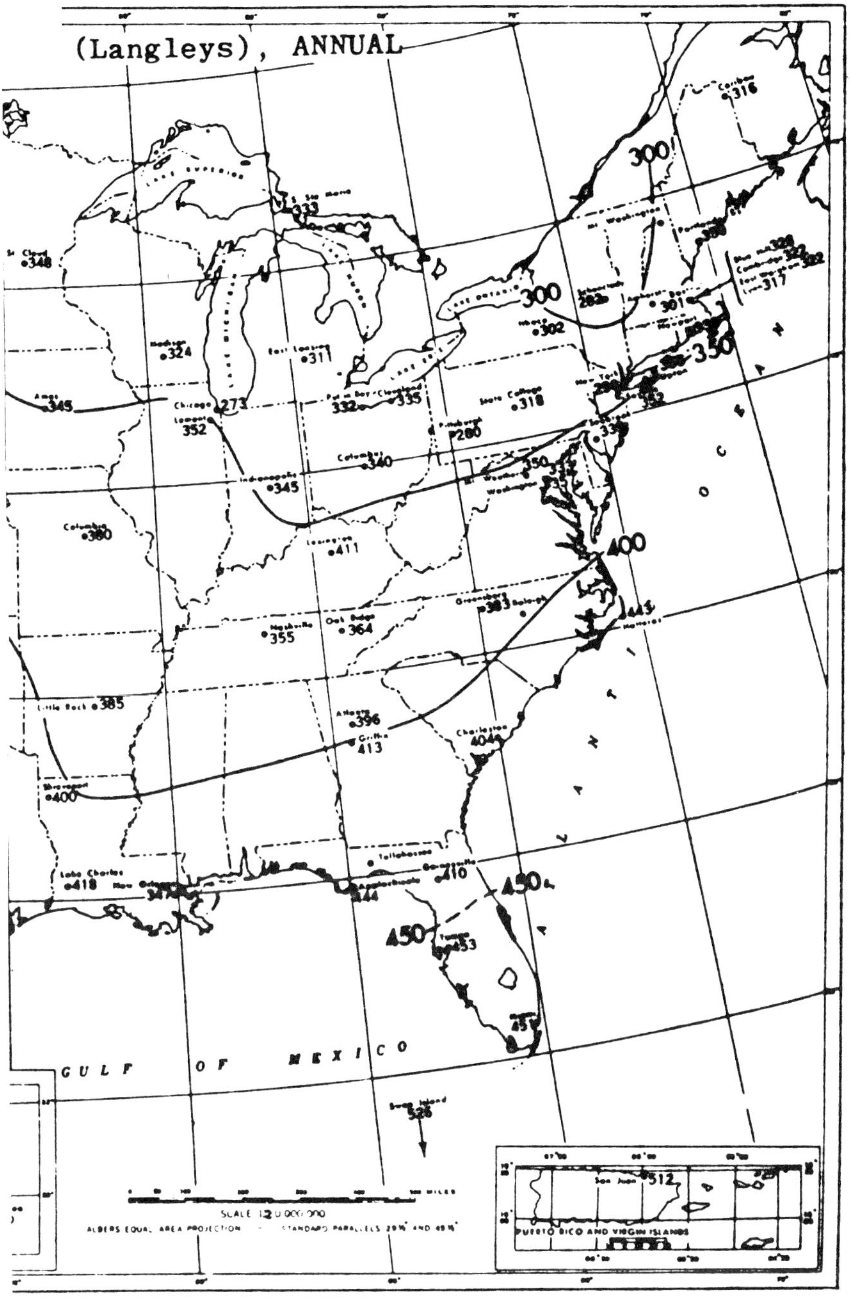
(Langleys), ANNUAL
300
300
350
400
450
450
GULF OF MEXICO
ATLANTIC OCEAN
PUERTO RICO AND VIRGIN ISLANDS
San Juan 512
SCALE 1:20,000,000
ALBERS EQUAL AREA PROJECTION — STANDARD PARALLELS 29½° AND 45½°

In 1913, near Cairo, Egypt, F. Shuman and C. V. Boys (engineers from Philadelphia) built a large solar engine of over 40 kilowatts. The long parabolic cylinders focused the solar radiation into a central pipe as shown in the figure. The steam drove an irrigation pump. (Courtesy of NASA.)

An experimental solar-energy-collector using a mirror system located at Odeillo, France, which achieves high temperatures. The absorber is located in the six-story building that appears in front of the mirror system. The parabolic mirror is about 30 meters high. (Courtesy of French Embassy Press and Information Division.)

ESTIMATES OF THE FUTURE CONTRIBUTION OF SOLAR ENERGY

Solar Technology	*1985*	*2000*	*2020*
Direct Thermal Applications (in Units of 10^{15} Btu = 1 Q per year)			
Heating and Cooling	0.15 Q	2.0 Q	15 Q
Agricultural Applications	0.03	0.6	3
Industrial Applications	0.02	0.4	2
TOTAL	0.2 Q	3 Q	20 Q
Solar Electric Capacity (in Units of 10^9 Watts = 1 Gwe			
Wind	1.0 Gwe	20 Gwe	60 Gwe
Photovoltaic	0.1	30	80
Solar Thermal	0.05	20	70
Ocean Thermal	0.1	10	40
TOTAL	1.3 Gwe	80 Gwe	250 Gwe
Equivalent Fuel Energy	0.07 Q	5 Q	15 Q
Fuels from Biomass	0.5 Q	3 Q	10 Q
Total Solar Energy	~1 Q	~10 Q	~45 Q
Projected U.S. Energy Demand	100 Q	150 Q	180 Q

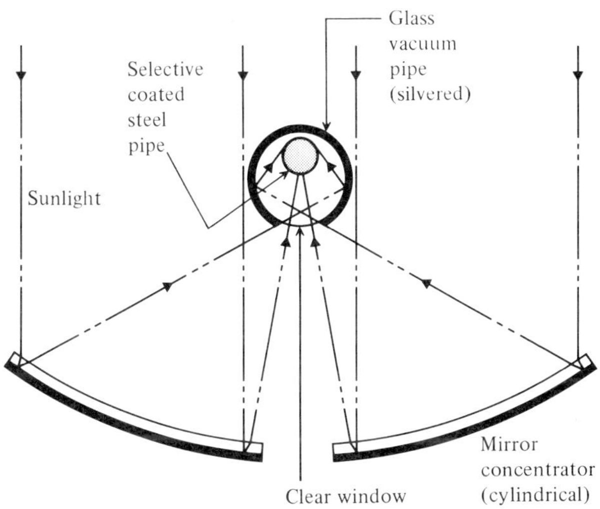

Cross section diagram of a typical collector using a cylindrical rear-surfaced mirror as the concentrator. The cylindrical mirror concentrates sunlight so that the focused beam enters an evacuated glass pipe through a transparent region. The remaining portion of the interior of the glass pipe is silvered to produce a high reflectivity. The sunlight entering this enclosure either impinges directly, or after a single reflection at the walls, on a steel pipe. The steel pipe is coated on its exterior with a selective coating. The enclosure is evacuated so that no heat is lost by convection. The result is that the steel pipe gets very hot, and if no way of extracting the absorbed energy is available, the pipe will rise in temperature to the order of 600°C. (*Source:* A. B. and M. Meinel, *Applied Solar Energy: An Introduction,* Addison-Wesley Publishing Co., 1976.)

PERCENTAGE OF HEATING AND COOLING THAT CAN BE READILY CONTRIBUTED BY SOLAR SYSTEMS

City	*Percent of Heating*	*Percent of Cooling*
Los Angeles	100%	98%
Atlanta	85%	63%
New York	84%	77%
Chicago	50%	67%

ECONOMIC COMPARISONS FOR A SOLAR SPACE AND HOT WATER SYSTEM FOR SINGLE-FAMILY DETACHED HOMES WITH AND WITHOUT THE FEDERAL TAX CREDIT*

	Electricity		*Fuel Oil*	
City and Criterion	*Without Credit*	*With Credit*	*Without Credit*	*With Credit*
Boston				
Payback Period (Years)	18	17	27	25
Years to Positive Cash Flows	3	3	13	13
Los Angeles				
Payback Period (Years)	13	11	Not Used	
Years to Positive Cash Flow	1	1		

*A tax credit is assumed as 30% of the first $1,500 expenditure and 20% of the next $8,500 of expenditures up to a maximum credit of $2,150. The solar system is competitive with electric heating, but not yet competitive with fuel oil (1979 prices). A solar water heating system (without space heating) is economically competitive in Los Angeles and Boston. *Source:* R. H. Bezdek, et al., "Economic Feasibility of Solar Water and Space Heating", *Science,* March 23, 1979, pp. 1214–1220.

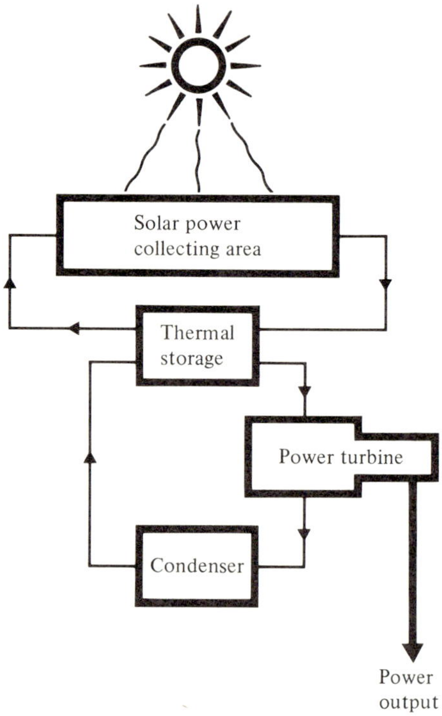

A schematic diagram of a solar thermal electric generating system. In this system, a heat transfer fluid or gas transfers the heat to a thermal storage tank until it is needed by the turbine.

The solar power tower facility at Sandia Laboratory in Albuquerque, New Mexico. The 140-foot tower accommodates test systems and the 62 sun-tracking mirrors focus the sun on the aperture in the tower.

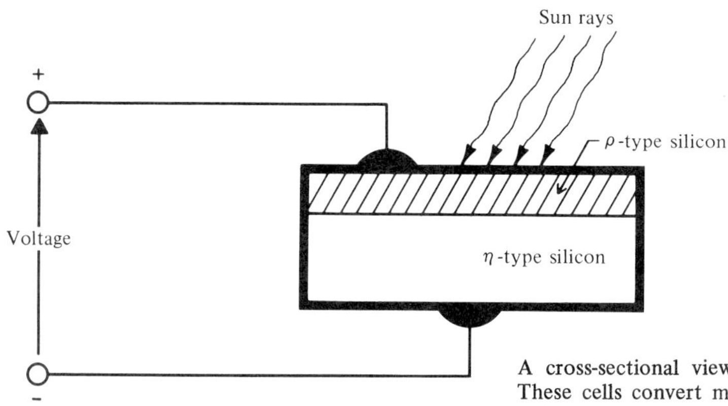

A cross-sectional view of a silicon solar cell. These cells convert more than 10% of the incident light energy into electricity.

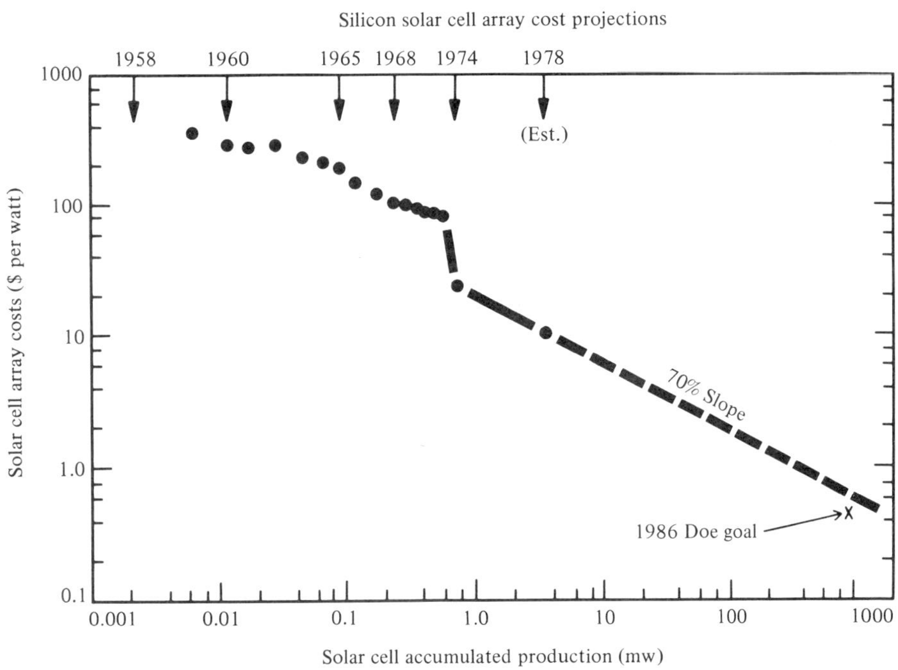

The history (solid line) and projection (dashed line) of the price of silicon solar cell arrays in 1976 dollars. The 70% learning curve is common in the semiconductor industry. (*Source:* D. Redfield, "Solar Energy and Conservation", *IEEE Industry Applications Society Record,* Sept. 1978, pp. 4–9.)

A Satellite Solar Power Station using concentrators and solar voltaic cells. The power is transmitted to earth by means of the microwave antenna located between the solar arrays. (Courtesy of Arthur D. Little, Inc.)

AVAILABILITY OF SOLAR ENERGY IN SYNCHRONOUS ORBIT FOR A SATELLITE SOLAR POWER STATION

Availability Factor	*Average on Earth*	*In Synchronous Orbit*	*Average Ratio, Ground/Orbit*
Solar Energy Density, kW/m^2	1.1	1.4	4/5
Hours of Useful Radiation/Day	8	24	1/3
Percentage of Clear Skies	50	100	1/2
Cosine of Angle of Incidence	0.5	1	1/2
Product of all Factors			1/15

PARAMETERS OF A 5000-MW SSPS*

Single Solar Panel Size	5.2 km X 4.33 km
Total SSPS Size	11.73 km X 4.33 km
Satellite Mass	10 X 10^6 kg
Transmitting Antenna Diameter	1 km
Receiving Antenna Diameter	7 km
Total System Cost	$10 Billion
System Cost per kW	$2,000

*SSPS = Satellite Solar Power Station

16

ENERGY STORAGE SYSTEMS AND ALTERNATE TECHNOLOGIES

Energy storage involves the collection and retention of readily-available energy for later use. Often sources of energy are available at periods of low demand and should be stored for times of peak demand. Often the peak power demand of an electric distribution system can be twice the minimum level of power demand. Thus, storage of hydroelectric, wind energy, or solar energy will be necessary in the last decades of this century.

In spite of the many advantages of electricity at the stage of fuel utilization, its generation and transmission creates many difficult problems. Electricity is difficult and costly to store for periods of peak demand and costly to transmit to the point of final use. Furthermore, the generation of electricity by the conversion of solar, wind, geothermal, nuclear, or thermal energy is a low-efficiency process. Electricity is a carrier of energy which has been converted from its original form in fossil fuels or the sun, for example, to electricity which can be transmitted over distances to cities or towns where it is used. The limitations of electricity cause consideration of other carriers such as hydrogen or methanol. Hydrogen has many desirable storage and transmission characteristics serving as an energy carrier.

Since solar energy is intermittently available, storage of the energy is desirable. A major cost of a solar thermal power plant is storage. Simple, short-term thermal-energy storage of 3 to 8 hours may comprise single-tank units using the working fluid of the system as a storage medium. Besides providing storage, the storage system provides thermal inertia in the receiver-loop carrying energy to the turbine generator, which is desirable to smooth out fluctuations in incident energy due to passage of clouds.

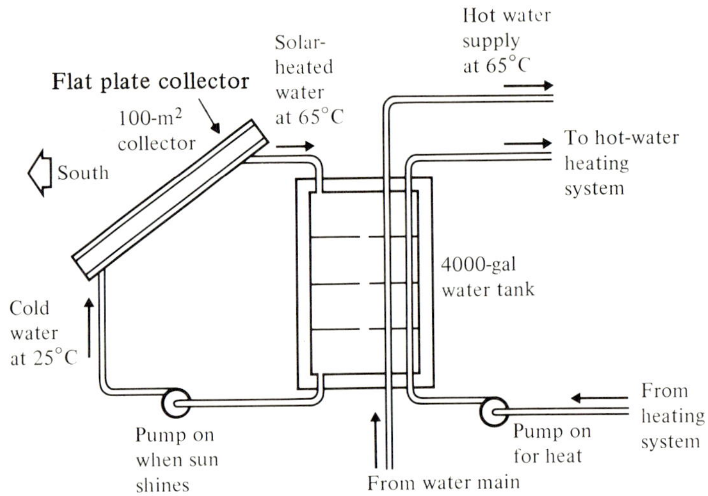

A schematic diagram of a system for space heating and hot water heating for a house of 1,800 ft.^{2}. The water tank provides 3 days heat storage.

The Decade 80 Solar house in Tucson, Arizona (1975). The house uses a combination copper roof and solar energy panel. About 1,800 sq. ft. of collectors are used to supply energy to a 3,000-gallon water-storage tank. Two lithium-bromide water-absorption units are used for cooling. (Courtesy of the Copper Development Association.)

ENERGY DENSITY OF ENERGY STORAGE SYSTEMS

	Pressure (psi)	*Temperature (°F)*	*Energy Density (Btu/ft³)*
Hot Rocks or Metals	—	300	10,000
Molten Salts	—	400	15,000
Water	15	212	9,000
Hydrogen Gas	1,000	60	18,500
Hydrogen Liquid	15	−425	200,000
Hydrogen Hydride	—	—	250,000
Methanol	—	—	430,000
Gasoline	—	—	830,000
Lead-Acid Batteries	—	—	10,000

CHARACTERISTICS OF BATTERY STORAGE SYSTEMS

Advantages	*Disadvantages*
Minimal Sitting Problems	Limited Lifetime
Can be Placed Near the Load	High Cost at Present
Reasonable Efficiency of Storage and Discharge = 70%	Extensive Materials Use

ENERGY AND POWER CAPACITIES OF STORAGE BATTERIES

Type	*Energy Storage (Watt-Hours/kg)*	*Power Capacity (Watts/kg)*
Lead-Acid	30	20
Nickel-Cadmium	40	240
Silver-Zinc	120	300
Zinc-Air	120	200
Sodium-Sulfur	200	220
Target	220	55

Several of the figures given are the potential capacity of the battery and are not available in commercial batteries.

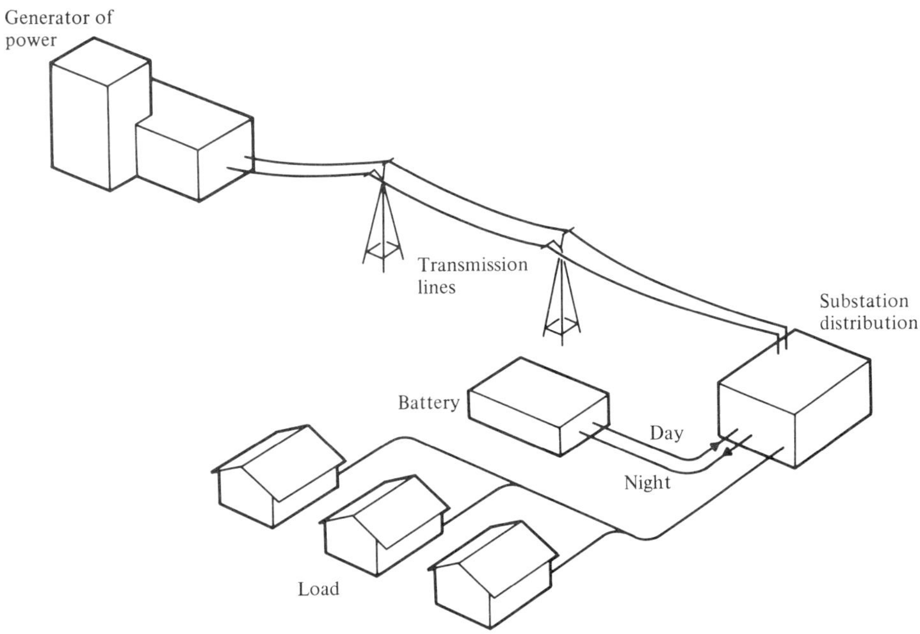

The generator of power is available all day while the load is low in the night. In this system the energy is stored in a battery system for peak-time use. For example, the generator could be a hydroelectric, a wind generator, or a nuclear power plant.

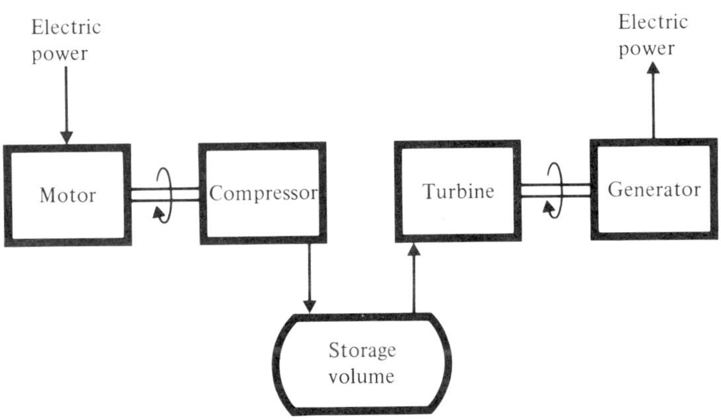

The basic schematic diagram of a constant-volume compressed-air energy-storage system. A compressed-air storage system uses an air compressor (pump) to place air in an underground cavity or a storage tank. When the stored energy is required, it is used to drive a turbine and an electric generator. During off-peak hours, energy is used to drive the pump and store air in the cavern or tank. Air can be stored in salt domes, mined caverns, depleted gas fields, abandoned mines, or an aquifer. The air is stored at 12 to 50 times atmospheric pressure. The overall efficiency of a compressed-air storage system is estimated to be 70%.

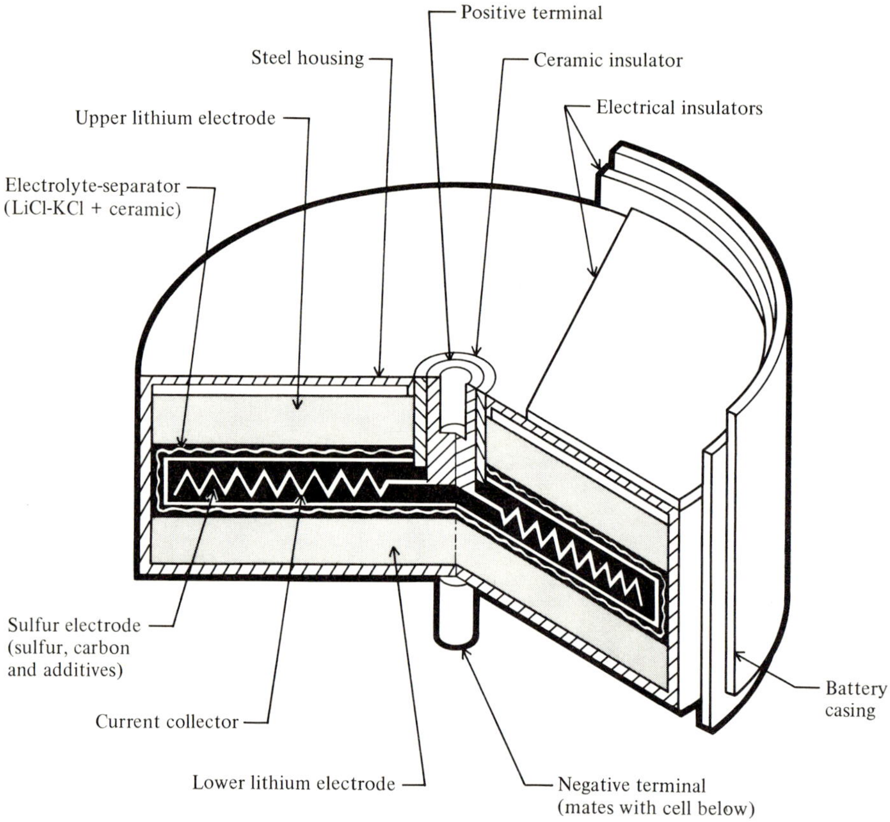

The lithium-sulfur battery cell produces a large amount of energy per unit weight and size. These cells can be stacked to propel an electric vehicle or for energy storage by a utility. For off-peak energy storage, large banks of batteries in an electric utility system would be charged during periods of low demand, and discharged during periods of high demand, thus leveling the load on the power-generating facilities.

MAXIMUM ENERGY STORAGE OF A FLYWHEEL

Material	*Tensile Strength* (10^6 *kgs/m*^{2})	*Density* (*gms/cm*^{3})	*Maximum Energy Storage* (*Watt-Hr/kg*)
Maraging Steel	281	8.0	55
E Glass	337	2.5	190
PRD-49	359	1.5	350
Fused Silica	1,406	2.1	870

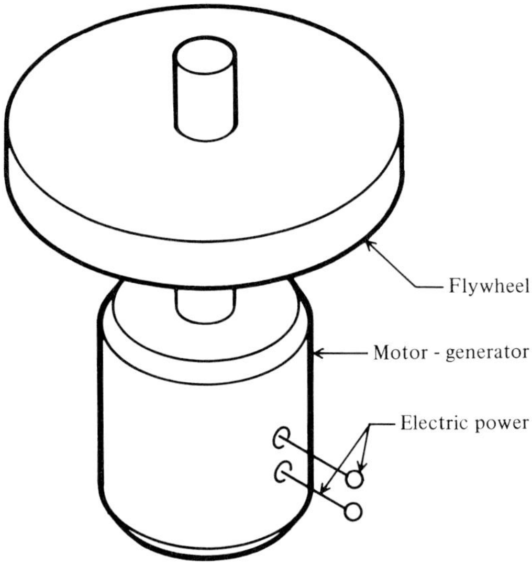

The flywheel, which is one of the oldest of human inventions, is an inertial storage device. The flywheel offers the potential for storage of energy on a large scale to handle the peak loads of utilities and to power electric vehicles. The maximum energy that can be stored in a flywheel disk is:

$$E_{max} = cVT$$

where V is the volume of the wheel, T is the tensile strength, and c is a constant (c = ½ for a solid disk).

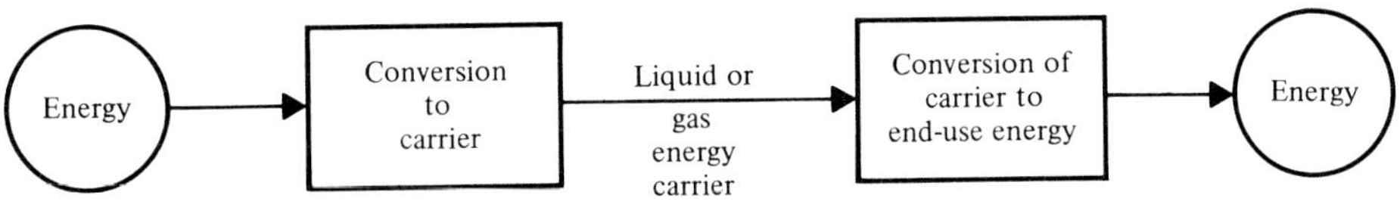

A schematic diagram of the desired energy carrier system. It is desired that a minimum of environmental pollution be generated at each conversion stage. The carrier must be able to readily store and transmit safely and at low cost.

PROJECTED COMPARISON OF THE CHARACTERISTICS OF ELECTRICITY AND HYDROGEN AT THE END OF THE 20TH CENTURY

Characteristic	*Electricity*	*Hydrogen*
1. Efficiency of Conversion to the Carrier	35 to 40%	35 to 50%
2. Transportation and Distribution	More Costly	Less Costly
3. Storage	Costly	Low Cost
4. Methods of Using the Carrier	Good in Industry Poor in Transportation	Good in Industry and Transportation
5. Safety	Proven Safe	Some Hazards

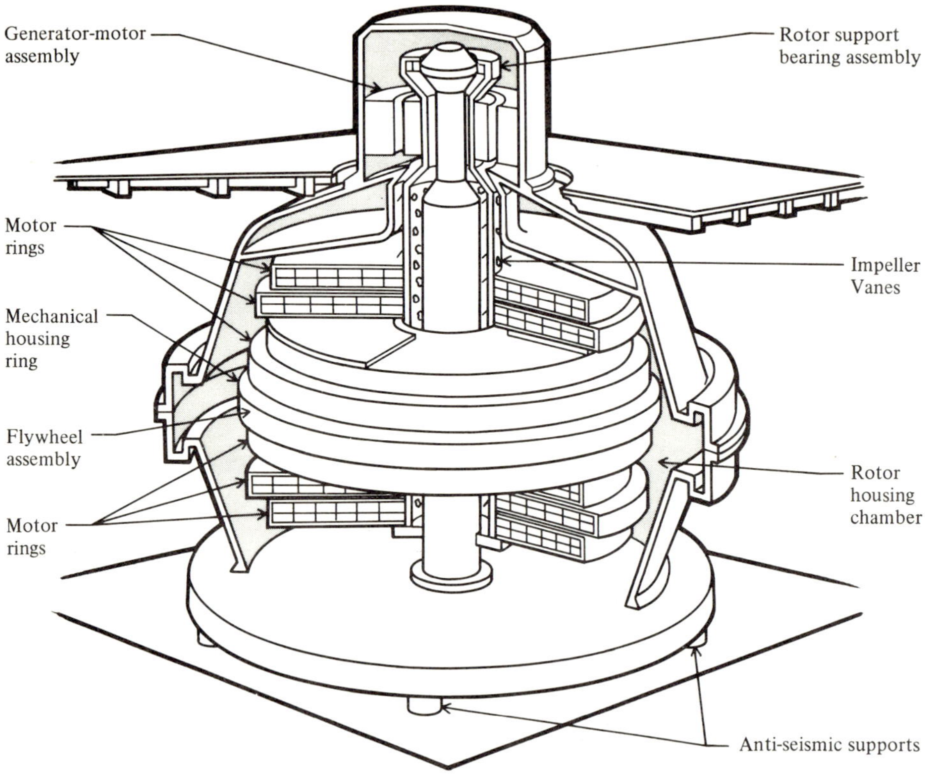

This peak-power unit would store energy from an electric power plant during off-peak hours and would generate energy at times of peak load. The fiber composite flywheel would be about 15 feet in diameter and weigh about 200 tons. It would be coupled to a generator-motor that would function as a motor when the flywheel was being spun up to store energy and as a generator when the system was drawing on the flywheel's stored energy. Each unit would store 10 megawatt-hours of energy at a rotation speed of 3500 revolutions per minute. The cost of such a system could be about $150 per kW.

CHARACTERISTICS OF A DESIRABLE ENERGY CARRIER

1. High efficiency of conversion of original energy source to the carrier fuel.
2. Availability of a method of transporting and disturbing the carrier fuel with low attendant energy losses and at low cost per unit distance.
3. Availability of a method of storing the carrier fuel for relatively long periods of time and at low cost.
4. Availability of several methods of utilizing the carriers fuel for transportation and industry.
5. Safe in handling and storage.

A carrier may be readily used to store and transmit energy. A gas or liquid is a natural candidate.

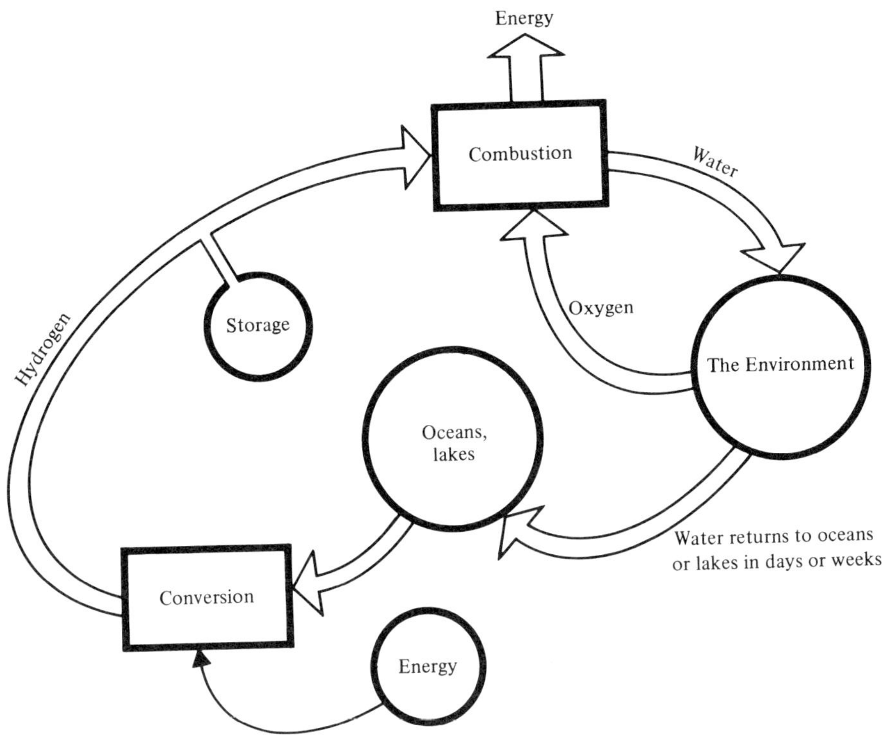

The hydrogen-conversion and end-use cycle. Hydrogen can be easily derived by the decomposition of water and is an excellent candidate for an energy carrier. Energy is converted to hydrogen by the thermal or electrical decomposition of water stored and transported to its end use. Hydrogen is readily used as a combustible fuel in combination with oxygen or air. The end product of the combustion is water which recycles to the oceans and lakes. There is a small amount of nitrogen oxide (NO) that is formed from the air entering the combustion, but it can be controlled. Otherwise, the hydrogen flame is free of pollutants.

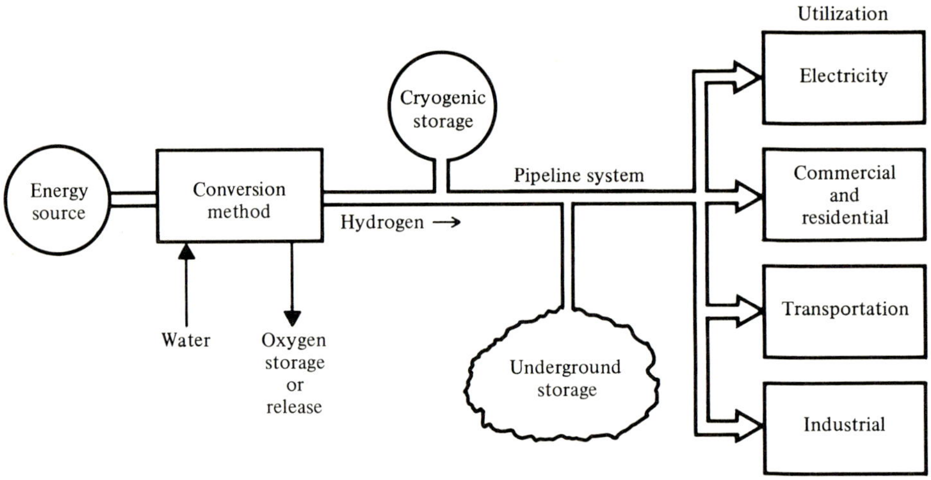

A storage and transmission system for hydrogen as an energy carrier. Hydrogen can be transmitted and distributed in liquid or gaseous form. In gaseous form, hydrogen can be transmitted in much the same way that natural gas is today – by means of pipelines. The movement of a gas by pipeline is one of the cheapest methods of energy transmission and a gas-delivery system is located underground where it is inconspicuous and safe. In the U.S. in 1970 there were 252,000 miles of trunk pipeline carrying a total of 22.4×10^{12} cubic feet of natural gas during the year. A similar pipeline system could be developed for hydrogen or the existing system could be used, in part, by hydrogen. A typical pipe is 36 or 48 inches in diameter and gas is pumped along the line by compressors spaced at 100 mile intervals. Pipelines operate at pressures of 600 psi (40 atmospheres) and are capable of power capacities of 10,000 megawatts, several times the capacity of an overhead electric transmission line.

ESTIMATED TRANSPORTATION COSTS PER 100 MILES OF THREE ENERGY CARRIERS

Transportation	*Cost (Dollars/10^6 Btu)*
Methane by Pipeline	0.030
Hydrogen by Pipeline	0.033
Electricity by High Voltage Overhead Transmission Lines	0.21

CHARACTERISTICS OF ENERGY CARRIERS

Carrier Characteristic	*Hydrogen Gas*	*Hydrogen Liquid*	*Methanol*	*Gasoline*	*Methane*
Energy Volume Density (Btu/ft^3)	32 325	200,000	430,000	830,000	1,025
Energy per Unit Mass (Btu/lb.)	58,000	58,000	9,200	19,400	23,000
Minimum Ignition Energy (mJ)	0.02	0.02	0.25	0.25	0.28
Ignition Temperature (°K)	858	858	783	530	810
Flammability Limits (% in Air)	4.1-75	—	2.2-9.5	1.5-7.6	5.3-15

ENERGY STORAGE DEVICES

Phenomena	*Energy Density (Watt-Hr/kg)*	*Power Density (Watt/kg)*	*Discharge Cycle Life (Cycles)*
Air Compression	7.7	10^4	10^7
Lead-Acid Battery	17.8	80	500
Nickel-Cadmium Battery	30.6	80	2,000
4,340 Steel Fly-Wheel	33.2	10^4	10^5
Maraging Steel Fly-Wheel	55.0	10^4	10^5

COSTS OF STORAGE SYSTEMS

Type	*Capital Costs (Dollars per kWh)*
Thermal Storage in Home Solar System (Water Tank)	\$ 4
Pumped-Hydroelectric	10
Lead Acid Batteries	50
Compressed Air Underground	50
Flywheels	150
Hydrogen Gas or Liquid	200

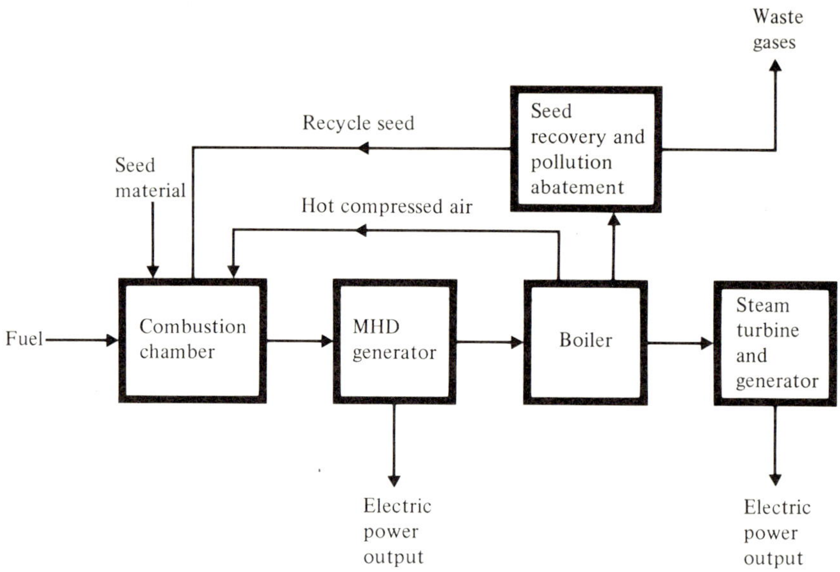

Magnetohydrodynamics (MHD) is the study of the flow of a conducting fluid in a magnetic field. The use of MHD electric power plants is possible by the end of this century if several technical problems can be solved. A schematic diagram of an electric power plant using MHD and a steam turbine is shown. The process relies on the passage of a hot, ionized gas under high pressure through a magnet coil. The source of the plasma gas can be petroleum, coal, or natural gas. The Russians are studying natural-gas units while the U.S. is exploring coal-fired units. The fuel is subjected to heat in the temperature of 2,500°C and ionizes and is capable of carrying an electric current. With such a high temperature, a theoretical efficiency of 80% could be achieved. In practice, an efficiency of 60% is expected, which is almost twice that normally achieved in a fossil-fuel boiler plant.

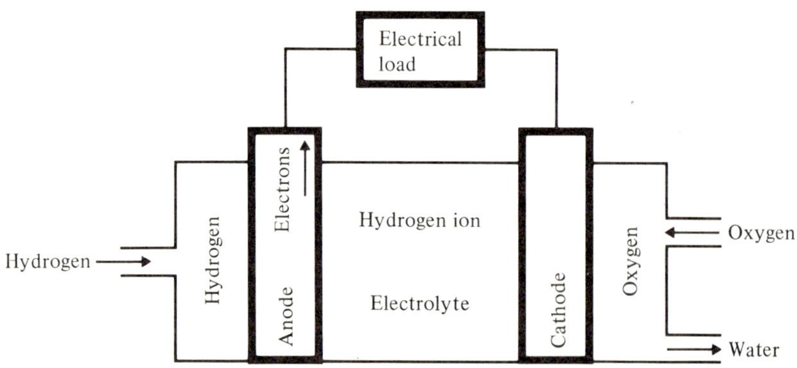

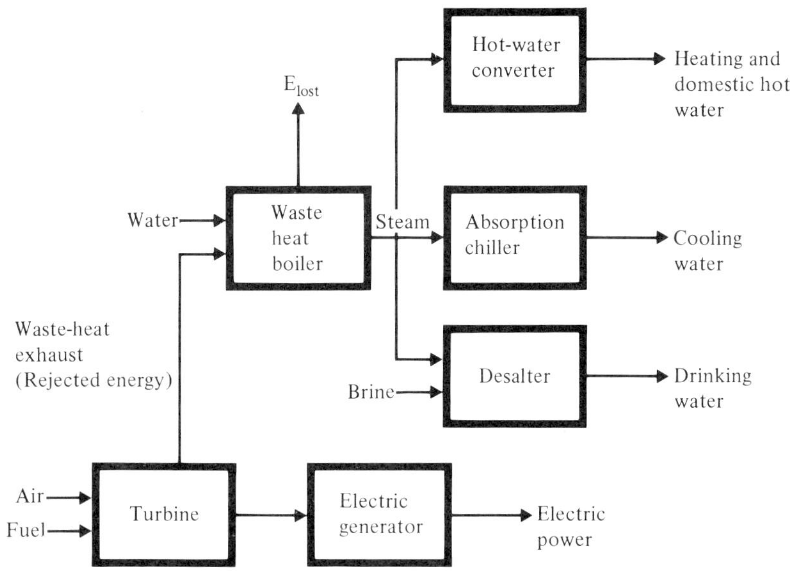

The fuel cell was first discovered by Sir William Grove in 1839, but the first practical cell was not demonstrated until 120 years later. While fuel cells are used in space vehicles for electric power, they have not been widely used due to high costs (presently $500/kW with a lifetime of several years). A single-fuel cell is a single set of two electrodes joined by electrolyte which conducts current between the electrodes and an external circuit. A schematic diagram of a hydrogen/oxygen fuel cell is shown. Hydrogen is fed to the anode where it forms hydrogen ions, H^+, by giving up electrons that travel through the external circuit. At the cathode, oxygen combines with hydrogen ions and electrons to form water. The overall efficiency of a hydrogen/oxygen fuel cell is about 60%. Each single fuel cell has a potential difference of about 1 volt so many cells are placed in series to yield a larger voltage and power output. Using 1,000 cells would yield a 1,000-volt array with a power output of 10 megawatts if each cell has a cross sectional area of 5 square meters.

A schematic diagram of a total-energy system. Within the next 10 years, total-energy systems may be developed and implemented on a sufficiently wide basis to become an important part of the U.S. energy system. *Total-energy* or *integrated-utility systems* are combined processing plants that generate electricity; use residual and recycled energy for heating, air conditioning, and hot water; treat water; process solid wastes, and treat liquid waters. The systems are often called *cogeneration systems.* Normally, about 65% of the fuel-energy content is wasted in the generation of electricity. In an integrated system, better than half of this "waste" energy can be recovered and used for space heating, air conditioning, domestic hot water, water and liquid waste treatment, etc. Similarly, a system required to provide process heat can also be used to generate electricity within the integrated system. The total or integrated system is located near the ultimate consumer in order to minimize the distribution costs and attendant energy losses.

17 CONSERVATION OF ENERGY

It is not only important to increase the supply of energy to meet the demands of the nations of the world, but it is equally important to reduce the demand for fossil fuels. The conservation of fossil fuels, and all forms of energy, can be achieved by an integrated, well-planned approach.

The conservation of energy can reduce the energy needs of the U.S. by a significant percentage. Estimates of ten- to forty-percent reduction in our overall energy consumption have been proposed as realistically achievable over the next decade. There are essentially four methods for energy conservation which are: (1) eliminate waste, (2) shift to less energy-intensive processes, (3) reduce energy-consuming activities, and (4) improve efficiency of energy-consuming activities.

Waste can be eliminated by simple actions such as repairing hot water leaks and broken windows and turning off unused lights. A shift to less energy-intensive processes may be illustrated by the use of mass transportation or the recycling of materials. A reduction in energy-consuming activities essentially implies a change of lifestyle less dependent upon automobiles, jet airplanes, and air-conditioned homes. The improvement of efficiency can be achieved in heating and cooling systems, industrial processes, etc.

THE ERDA* CONSERVATION PLAN OF 1975

Energy Saving Source	Annual Energy Savings (10^{15} Btu) 1985	2000
Conversion of Waste Materials	2.5	4.5-9
Electric Conversion Efficiency Improvement	0	4.5
Electric Power Transmission and Distribution	0	4.5
Electric Transportation	0	4.5
Transportation Efficiency Improvement	2.5-6	4.5-9
Industrial Energy Efficiency	2.5-6	4.5-9
Conservation in Buildings and Consumer Products	2.5-6	4.5-9
TOTAL	10-27	31.5-49.5

*ERDA = Energy Research and Development Administration now merged into the U.S. Department of Energy.

ENERGY CONSERVATION SCHEMES

	Potential Near Term Savings as Percent of Total for that Item
Residential	
a. Insulation (homes)	5%
b. Air Conditioning	1%
c. Lighting, Appliances	5%
d. Solar Heating	5%
Industry	
a. Process Steam	5-10%
b. Recycling of Materials	3%
c. Total Energy Scheme	5-10%
Transportation	
a. Mass Transit for 10% of Urban Passenger-Miles	2- 5%
b. Smaller, More Efficient Autos	3- 5%
c. Shift Some Freight and Passengers to Trains	1- 2%
Commercial	
a. Air Conditioning	2%
b. Solar Heating	5%
c. Building Design	3%

FIRST LAW ENERGY EFFICIENCIES OF SEVERAL PROCESSES

Device	*Efficiency, η (Percent)*
Electric Generator	98
Electric Motor	90
Steam Boiler	88
Home Gas Furnace	85
Home Oil Furnace	65
Steam-Electric Power Plant	40
Diesel Engine	38
Automobile Engine	25
Fluorescent Lamp	20
Photovoltaic Cell	10
Incandescent Lamp	5

First Law efficiency is defined as

$$\eta = \frac{\text{energy output}}{\text{energy input}}$$

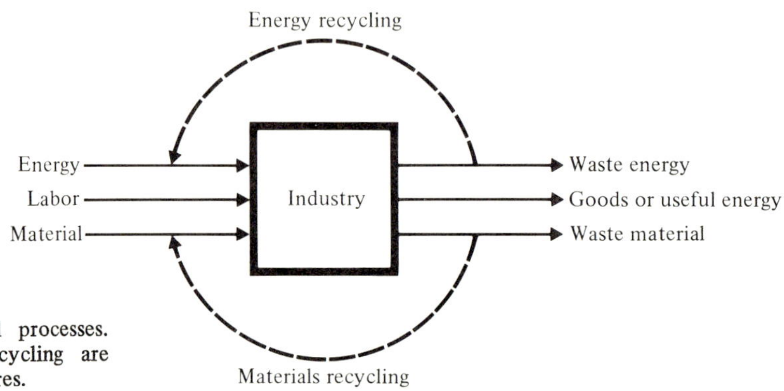

The use of energy in industrial processes. Energy-recycling and materials-recycling are both important conservation measures.

SECOND LAW EFFICIENCY OF SEVERAL PROCESSES

	Relative Thermodynamic Quality	*Percent of U.S. Consumption (1968)*	*Second Law Efficiency*
Use			
Space Heating	Lowest	18	0.06
Water Heating	Low	4	0.03
Air Conditioning	Lowest	2.5	0.05
Industrial Uses			
a. Process Steam	Low	17	0.25
b. Direct Heat	High	11	0.30
c. Electric Drive	High	8	0.30
Transportation			
a. Automobile	High	13	0.10
b. Truck	High	5	0.10

Second law efficiency is defined as

$$\epsilon = \frac{\text{theoretical minimum energy required}}{\text{energy actually consumed}} = \frac{B_{\text{min}}}{B_{\text{act}}}$$

where B is available work. The amount of work needed to heat a home is small compared with the amount of work that would be extracted from combustion of fuels. By contrast, these fuels could be used to generate high temperature differences in modern plants and use more of the available work. For the transfer of thermal energy Q to a reservoir (for example, a room being heated) at temperature T, we have

$$\epsilon = \frac{B_{\text{min}}}{B_{\text{act}}} = \eta\,(1 - T_o/T)$$

where T_o = ambient environment temperature (°K), and η is the first law efficiency.

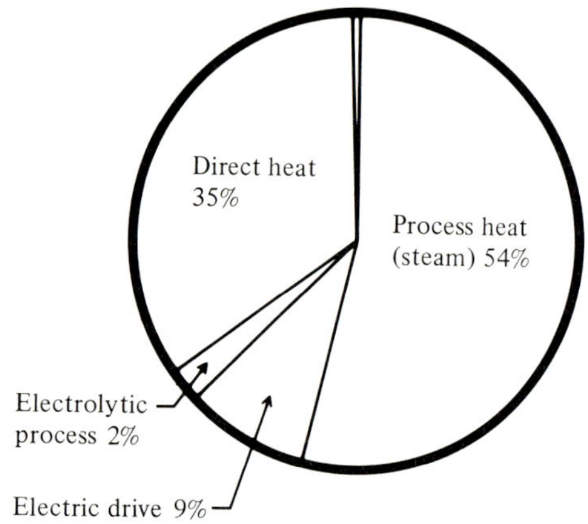

The primary uses of energy in industry. Conservation measures such as insulating and maintaining steam-pipe lines and improved efficiency of the use of direct heat are important.

INDUSTRIAL ENERGY CONSUMPTION

Industry or Product	*% of Industrial Energy Consumption*	*% of Total U.S. Energy Consumption*
Iron and Steel	13.6%	5.6%
Petroleum Refining	11.3	4.6
Paper and Paperboard	5.2	2.1
Petrochemical Feedstock	4.9	2.0
Aluminum	2.8	1.1
Cement	2.1	0.9
Ammonia	2.0	0.8
Ferrous Foundries	2.0	0.8

EFFICACY OF ELECTRIC LAMPS

Type	*Efficacy (Lumens/Watt)*
Low Pressure Sodium	180
High Pressure Sodium	130
Metal-Halide	100
Fluorescent	80
Mercury-Vapor	56
Incandescent	23

Efficacy = light output in lumens per watt input.

THE ENERGY INTENSITY OF SELECTED INDUSTRIAL PROCESSES

Product	*Energy Intensity* (10^6 *Joules/Dollar U.S.*)
Primary Aluminum	409
Plastics	229
Man-Made Fibers	214
Airlines	202
Paper Mills	187
Fertilizers	183
Glass	109
Fabricated Metal Products	97
Motor Vehicles	71
Wearing Apparel	53
Computers	38
Banking	19
U.S. AVERAGE	84

Energy Intensity = Energy used per dollar of value of the product.

ENERGY REQUIREMENTS OF ELECTRIC APPLIANCES

Appliance	*Average Power* (*Watts*)	*Average Annual Energy Consumption* (*kw-hr*)
Water Heater	2,500	5,000
Refrigerator (14 cu. ft. Frostless)	215	1,800
Freezer (15 cu. ft.)	341	1,200
Range (with Oven)	12,000	1,200
Dishwasher	1,200	360
Color Television	200	440
Air Conditioner		
Room	860	800
Central	1,300	2,000

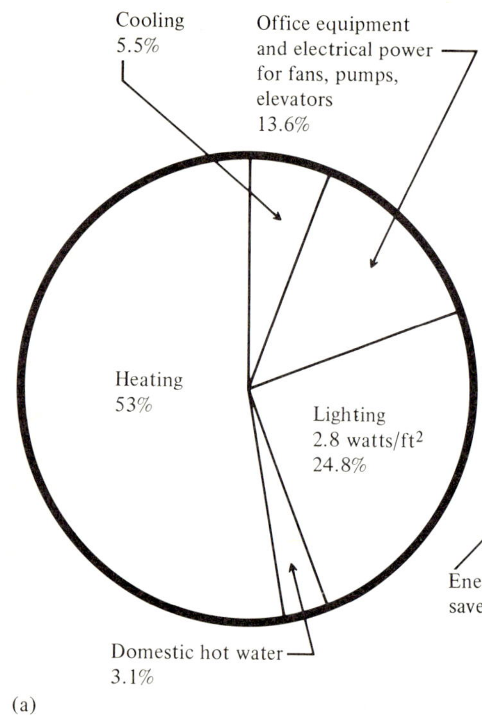

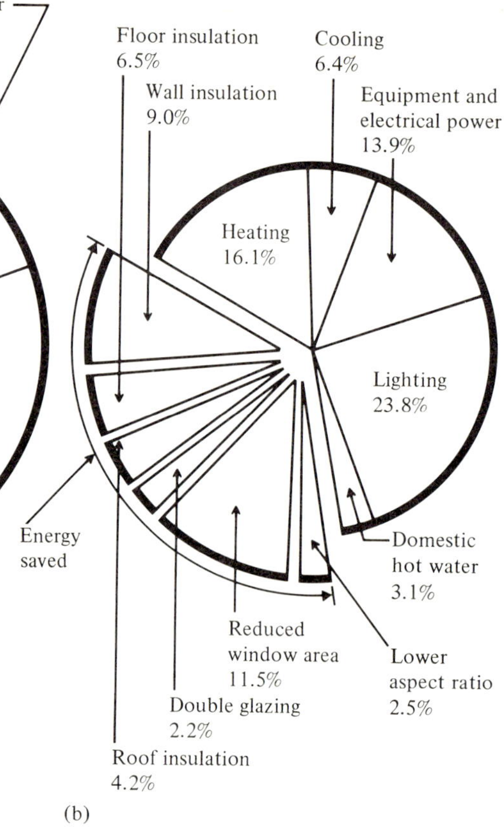

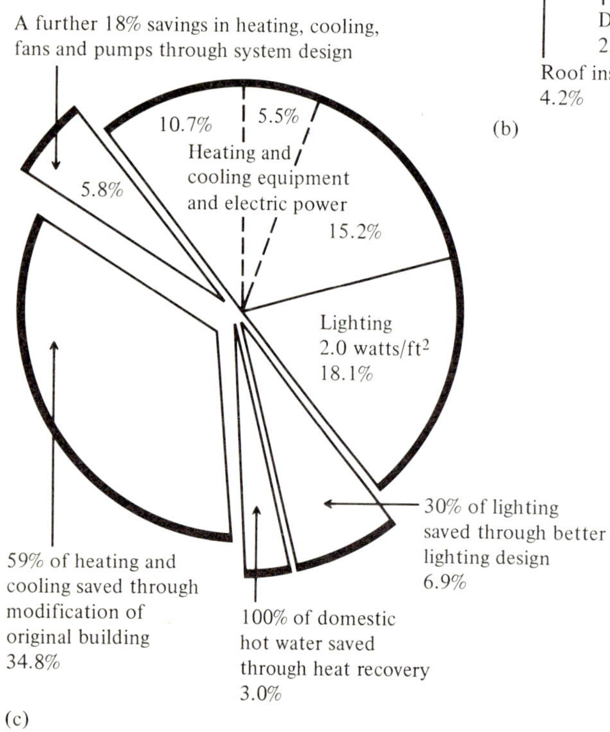

(a) Energy consumed in a typical 126,000 ft.^{2} office building in Manchester, N.H. is 13×10^9 Btu/year (14×10^{12} joules/year) (b) Improved design consisting of increasing wall, floor, and roof insulation reducing window area from 50 to 10%; adding double glazing; and changing the length/width of the building to 1/1 yielding a savings of 34%. (c) Improved lighting and heating, ventilating, and air conditioning systems cut another 16% off the annual energy consumption of the typical building. The Manchester building cost about 10% more but will consume about one-half as much energy as a typical building. (*Source:* G. Dallaire, "Designing Energy-Conserving Buildings," *Civil Engineering,* April 1974, pp. 54–58.)

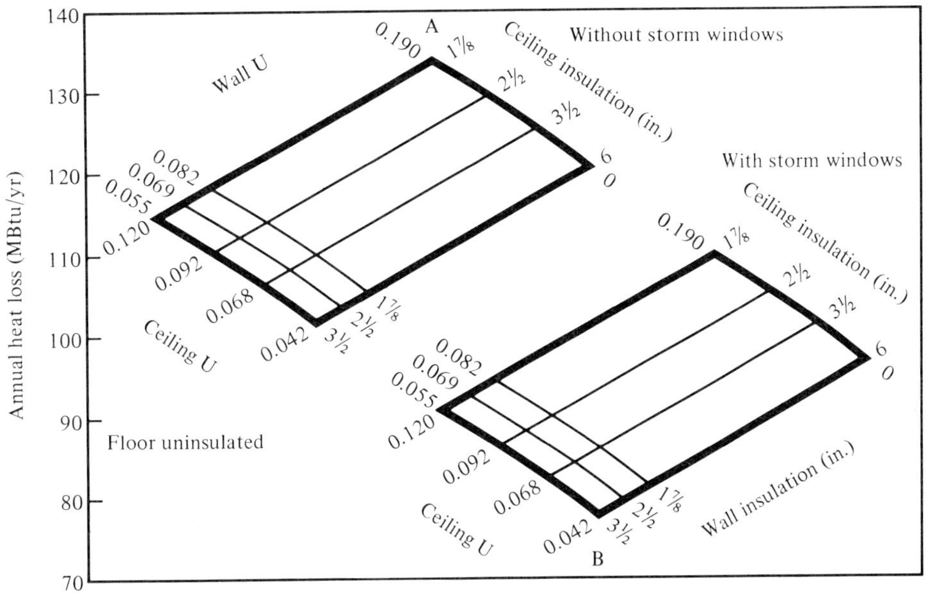

The annual heat loss from a typical home in New York is 10^6 Btu per year (U = Btu/ft^{2}hr°F). The annual heat loss is a function of ceiling and wall insulation as well as the use of storm windows (or dual glazing). The upper point A may represent most of the housing built prior to 1970. If storm windows and insulation are added so that the home operates at point B the heat loss is reduced from 134 to 78 × 10^6 Btu/year. (*Source:* C. A. Berg, "Energy Conservation Through Effective Utilization," *Science,* July 13, 1973, pp. 128–137.)

ENERGY CONTENT OF MATERIALS AND MANUFACTURED PRODUCTS

	Energy (Megajoules/kg)	*Cost of Energy/Value of Product*
Metals		
Steel	40	0.3
Aluminum	100	0.4
Copper	30	0.05
Magnesium	90	0.1
Products		
Glass Bottles	40	0.3
Plastic	10	0.04
Paper	25	0.3
Inorganic Chemicals	12	0.2

ENERGY BALANCE IN A STANDARD HOUSE AND IN A TARGET HOUSE

	Power in Btu/Hr			
	Standard House		*Target House*	
	Mild, Clear Day (ΔT=30°F)	*Cold, Partly Cloudy Day (ΔT=70°F)*	*Mild, Clear Day (ΔT=30°F)*	*Cold, Partly Cloudy Day (ΔT=70°F)*
Energy Losses				
Walls and Roof	7,200	16,800	2,800	6,500
Floor	1,100	1,100	400	400
Windows	6,600	15,400	1,800	4,200
Incoming Air (Heating)	8,000	18,700	1,600	3,700
Incoming Air (Humidification)	3,800	3,800	800	800
	26,700	55,800	7,400	15,600
Energy Gains				
Solar Through Windows	5,200	2,600	4,000	4,200
3 People	1,000	1,000	1,000	1,000
Electric Load	5,700	5,700	2,400	2,400
Gas Furnace Load to Balance	14,800	46,500	—	8,000
TOTAL	26,700	55,800	7,400	15,600

About 8×10^{15} Btu was consumed in the U.S. in 1978 for residential space heating. For a two story home with dimensions 25 $\times$ 30 $\times$ 20 feet, consider the target house with increased insulation and reduced infiltration.

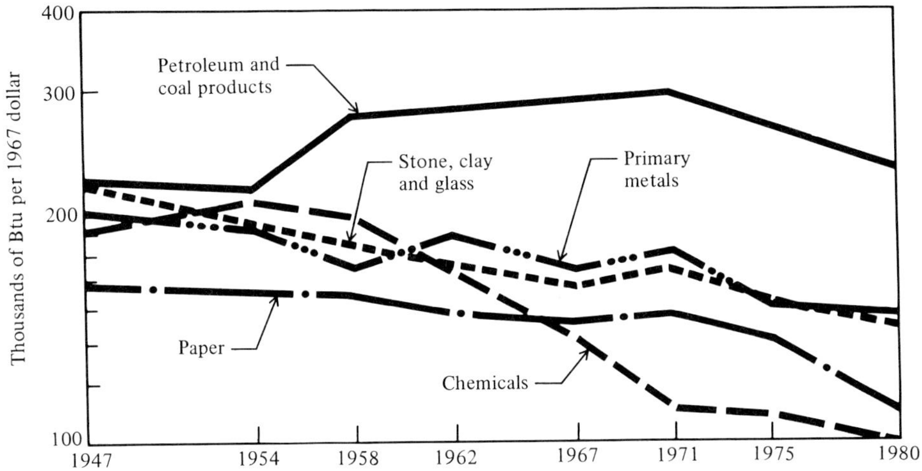

Energy consumption in relation to product value added during manufacture. (*Source:* The Conference Board, *Energy Consumption in Manufacturing,* 1974 Ballinger Publishing Company, Cambridge, Mass.)

ENERGY SAVINGS FROM RECYCLING MATERIALS IN THE U.S.

Material	*kWh Saved/Ton of Recycled Material*	*Estimated Total Savings if All Material was Recycled (kWh)*
Paper	4,210	50×10^9
Steel for Autos	7,000	100×10^9
Aluminum Beverage Cans	49,000	7.2×10^9
Steel Beverage Cans	7,000	34×10^9
TOTAL		191.2×10^9
Total Electrical Energy Use in the U.S. in 1975		$1{,}900 \times 10^9$

RECYCLING OF MATERIALS

Material	*Point of Impact*	*Energy Cost (kWh/Ton)* Virgin Material	Recycled Material	*Percent Saved*	*Ease of Recycling*
Steel	Molten Steel	13,680	6,636	52	Present Technology Limited by Impurities, Separation, and Collection.
Aluminum	Molten Aluminum	65,780	2,400	96	Present Technology Limited by Separation and Collection
Plastics	Molten Polymer	13,238	586	96	Extremely Difficult; No Satisfactory Technology Known.
Paperboard	Pulp	1,923	970	50	Present Technology for Separated Waste; No Satisfactory Technology for Mixed Waste.
Glass	Transportation	2,287	2,287	0	Difficult and Costly.

Source: L. Schipper, "Towards More Productive Energy Utilization," Lawrence Berkeley Laboratory Report 3299, Oct., 1975.

ENERGY INTENSITY OF SELECTED INDUSTRIES PER EMPLOYED PERSON

Industry	*Energy Intensity per Employed Person (10^{15} Btu per Employee)*
Chemical	14.0
Automobile	8.0
Paper	7.0
Construction	5.5
Food	3.0
Electrical	3.0
Wood	2.0

A SET OF OBJECTIVES FOR ENERGY CONSERVATION FOR THE U.S. IN 1985

Item	*Annual Energy Saving (Quad = 10^{15} Btu)*
1. Use of Wastes for Recycling and Energy Generation	2.0
2. Electric Conversion and Transmission Efficiency Improvement	0.5
3. Use of Waste Heat from Electric Plants for District Heating	1.0
4. Efficiency Labelling and Efficiency Improvement for Appliances and Room Air Conditioners	0.5
5. All New Automobiles with at Least 22 Miles per Gallon Average	2.0
6. Industrial Energy Efficiency Improvement	3.0
7. Conservation in Building, Heating and Cooling	2.0
8. Revision of Peak Load Pricing Schemes for Electric Utilities	1.0
TOTAL	12.0

PROGRESS TOWARDS ACHIEVING ENERGY CONSERVATION GOALS

Industry	*Percent Improvement in Energy Efficiency*	
	Realized 1976	*Target 1980*
Food	11	12
Chemicals	10	14
Primary Metals	4	9
Machinery	19	15
Stone, Clay, Glass	8	16
Petroleum and Coal Products	10	12
Fabricated Metals	8	24

Source: U.S. Dept. of Energy, Annual Report, 1978, Vol. 1.

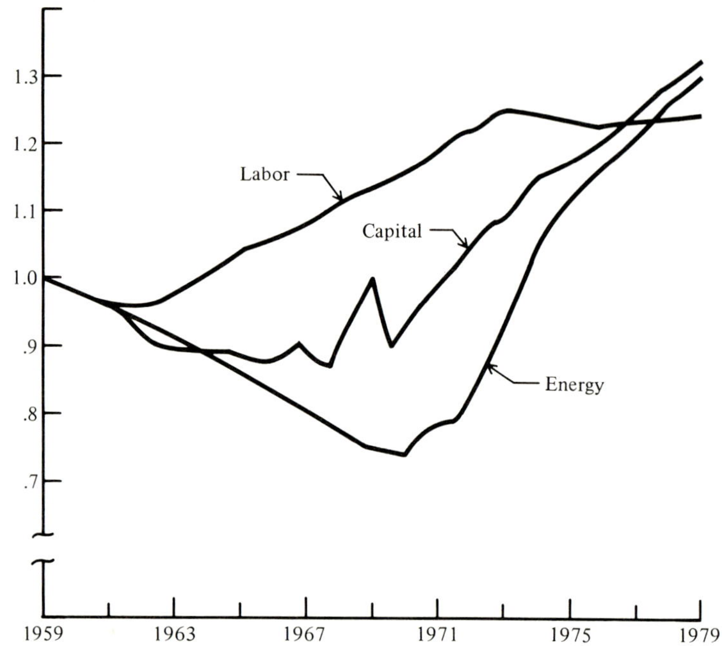

The real price of energy, capital, and labor from 1959 to 1979 (1960 = 1.0). Note that energy was a cheap ingredient in 1970 but has risen to equal labor and capital.

18

ENERGY AND THE ENVIRONMENT

Energy, the economy, and the environment are all interrelated. The environmental consequences of energy consumption have become matters of general concern only within the past two decades.

The demand for clean, nonpolluting energy has caused major shifts in the utilization of fuels. The electric utilities, under pressure to reduce sulfur dioxide pollution, have turned from coal to oil, thus causing an increase in oil consumption and eventually imports. Now, in order to avoid dependence on imported oil, the President has instituted a program requiring a shift back to coal.

In this chapter we will consider some overarching issues of energy and the environment. In each preceding chapter on a specific fuel or energy source, we reflected upon the environmental impacts of that fuel. Here we will attempt to examine some of the issues common to all fuels.

The use of energy eventually results in waste, heat, and matter. Environmental concerns relate to air pollution, water pollution, thermal waste heat, climatic changes of the earth, among others. Environmental problems are created at each stage of the production, upgrading, transportation, and utilization of fuels.

ENVIRONMENTAL FACTORS RELATED TO ENERGY

Environmental element affected	Type of impact	Pollutant	Areal extent: Global	Areal extent: Regional	Areal extent: Local	Cost of control: High	Cost of control: Intermediate	Cost of control: Low
Air	Chemical–physical	Carbon dioxide	■					
		Carbon monoxide			■	■		
		Sulfur oxides		■	■		■	
		Nitrogen oxides		■		■		
		Hydrocarbons		■				■
		Photochemical smog		■				
		Particulates	■	■	■		■	
		Water vapor	■		■			■
		Trace metals		■	■			■
	Radiological	Noble gases	■					■
		Particulates		■	■			
	Resource use	Oxygen						
		Helium		■				■
	Other	Thermal inputs		■	■			
		Electromagnetic emissions			■			■
		Noise		■	■	■		
Water	Chemical–physical	Oil spills		■	■			■
		Acid mine drainage		■	■	■		
	Radiological	Tritium	■					■
		Other effluents		■				■
		Uranium milling wastes			■			■
	Thermal	Thermal inputs		■	■		■	
Land	Chemical–physical	Acid fallout from the air		■	■			
		Mineral fallout from cooling towers			■			■
		Solid wastes			■		■	
	Radiological	High-level wastes		■	■		■	
	Resource use	Land subsidence			■		■	
		Strip-mining of coal		■	■	■		
		Land use for power production and transmission		■	■		■	
		Hydroelectric dams		■	■			

Costs: High – greater than $1 billion per year.
Intermediate – $\$10^6 < \text{cost} < \$10^9$ per year.
Low – Less than $1 million per year.

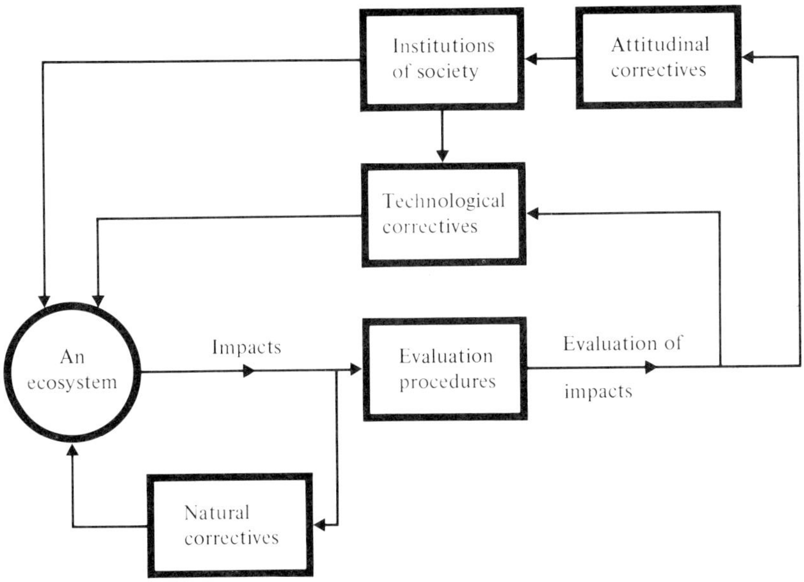

A model of the control of environmental and social impacts of an ecosystem. The impacts resulting from the interaction of energy use, man, and his environment are evaluated. These evaluations or estimates of the impacts result in attitudinal and technological corrective actions. This feedback process continues throughout time, and as people's attitudes are changed, they alter their institutions in order to improve the situation.

U.S. STACK EMISSION STANDARDS FOR PLANTS BUILT AFTER 1971

	Maximum Permissible Emissions in Lbs./Btu 10^6		
		Type of Fuel	
Pollutant	*Coal*	*Oil*	*Gas*
Particulates	0.1	0.1	0.1
Sulfur Dioxide	1.2	0.8	–
Nitrogen Oxides	0.7	0.3	0.2

These U.S. requirements for stack emissions from fossil fuel burners became effective in 1977. Coal normally releases about 5 lbs of SO_2 per 10^6 Btu in contrast to the requirement of 1.2 lbs/10^6 Btu.

Flue gas desulferization systems have been implemented in many power plants. Other plants use tall stacks and low-sulfur fuel. It is not yet clear when all the desired air quality standards can or will be met. Nevertheless, air quality schemes are being pursued vigorously by both government and industry.

COMPARATIVE COSTS OF COOLING SCHEMES FOR A 800 MEGAWATT FOSSIL FUEL POWER PLANT

	Cost (Millions of Dollars)
Once Through Cooling (River or Lake)	5
Cooling Pond	7
Wet Tower	
Forced Draft	7
Natural Draft	9
Dry Tower	
Forced Draft	16
Natural Draft	32

Cooling towers are classified as wet or dry and each of these may be forced draft or natural draft. In the wet tower, the cooling water is in contact with the air and cooling takes place by evaporation. Wet cooling towers operate well in cool, dry climates. Because of the water placed into the air, fogging or local rain can take place.

A dry tower is basically a large radiator. The cooling water flows through finned tubes and is cooled by a passing over them. Dry towers are more expensive to construct than wet towers but have no water loss. These costs can be compared to the cost of the plant which might average 400 million dollars.

COSTS OF POLLUTION CONTROL FOR ENERGY SYSTEMS

Total Cumulative Cost for 1971-1980 in 1971 Dollars	$287 Billion
Cost of Pollution Control Device on One Automobile in 1979	$400

EMISSIONS OF SULFUR DIOXIDE IN THE U.S. IN MILLIONS OF TONS PER YEAR

Year	1940	1950	1960	1970	1980 (Projected)
Emissions	22.8	24.9	23.2	33.4	30

The amount of SO_2 emitted in future years will depend upon the utilization of coal for fossil fueled power plants and the installation of scrubbers.

ANNUAL EMISSIONS OF A 1,000 MEGAWATT COAL-FIRED POWER PLANT USING EASTERN U.S. COAL

Gaseous and Liquid	
Carbon Monoxide (Tons/Year)	640
Carbon Dioxide (Tons/Year)	6,710,000
Sulfur Oxides (Tons/Year)	
No Controls	111,000
Wet Limestone Scrubbers	15,000
Nitrogen Oxides (Tons/Year)	
No Controls	20,300
Wet Limestone Scrubbers	17,900
Released Particulates (Tons/Year)	
No Controls	45,200
Precipitator	5,226
Solid	
Fly Ash or Slag (Tons/Year)	
No Controls	280,000
Wet Limestone Scrubber	897,000
Thermal	
Waste Heat (Megawatts)	
No Controls	1,560
Wet Limestone Scrubbers	1,860

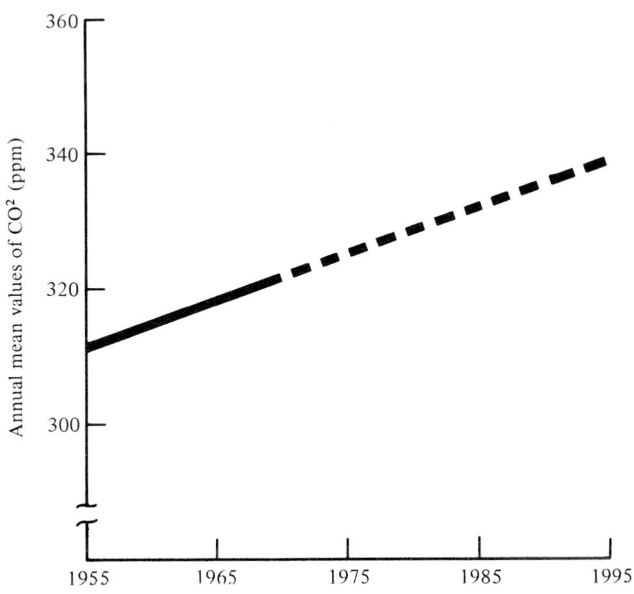

During the past decades, the combustion of fossil fuels has increased the atmospheric concentration of carbon monoxide from 300 parts per million (ppm) to 320 ppm in 1970. The dotted line projects the increase of the mean value of CO_2 in the atmosphere. A debate continues regarding the effects on the earth's climate of this increase of CO_2 in the atmosphere.

EMISSIONS OF POLLUTANTS BY SOURCE IN THE U.S. IN 1974 IN MILLIONS OF TONS PER YEAR

Source	*Particulates*	*Sulfur Oxides*	*Carbon Monoxide*	*Hydrocarbons*	*Nitrogen Oxides*
Transportation	1.3	0.8	73.5	12.8	10.7
Fuel Combustions in Boilers	5.9	24.3	0.9	1.7	11.0
Industrial Processes	11.0	6.2	12.7	3.1	0.6
Solid Waste Disposal	0.5	0.0	2.4	0.6	0.1
Other	0.8	0.1	5.1	12.2	0.1
TOTAL	19.5	31.4	94.6	30.4	22.5

Source: Environmental Protection Agency, 1976.

NATURAL AND MAN-MADE ENERGY DENSITIES

	Watts/m^2
Sun's Energy Absorbed on the Earth's Surface (Average)	100
Solar Energy Absorption by Atmosphere	25
Cooling Pond for Power Plant	100
Energy Consumption Density	
Moscow	127
Los Angeles	21
New York	600

DIRECT LOSSES OF CRUDE OIL INTO THE WORLD'S OCEANS IN 1980 IN METRIC TONS PER YEAR

Tankers	
Total Transported by Tanker	2.5 Billion Tons/Year
Loss by Accidents	200,000 Tons/Year
Loss Through Normal Operation	600,000 Tons/Year
Offshore Production	
Loss Through Normal Operations and Blowouts	300,000 Tons/Year
Refineries	
Loss Through Effluents to Oceans	400,000 Tons/Year
Total Loss into the Oceans	2,000,000 Tons/Year
Total Production in 1980	4,000 Million Tons/Year
Percent Lost	5%

19 ECONOMICS

The economics of energy concerns the allocation of the scarce resource, energy, which has various alternative uses. In part, prices are used to allocate this resource. Also, governmental policies, taxation, and trade quotas among others, are used to allocate the various alternative sources to potential consumers.

The economics of energy refers to its aspect as a commodity. Exchange takes place when energy is transferred from one party to another and some valuable good, usually money, is transferred in the opposite direction. Energy is mainly used as an input to an economic process and as an intermediate good. When energy is exchanged, it has a price in dollars per unit.

The affluence of a society is increased up to a point by the use of energy. Thus, there is a correlation between the Gross National Product (GNP) of a country and its input of energy during the nation's development phase. After affluence is obtained, then rising energy use will result from lifestyle and affluence characteristics themselves.

A secondary factor regarding energy use is its availability. The location of the source relative to the consumer results in a transportation problem. Also, there is a factor of time, that is, when this energy is available. The transportation and sale of energy requires some appropriate form of organization. Therefore, the existence of markets, pertinent regulation, and adequate organization are important parts of the economics of energy.

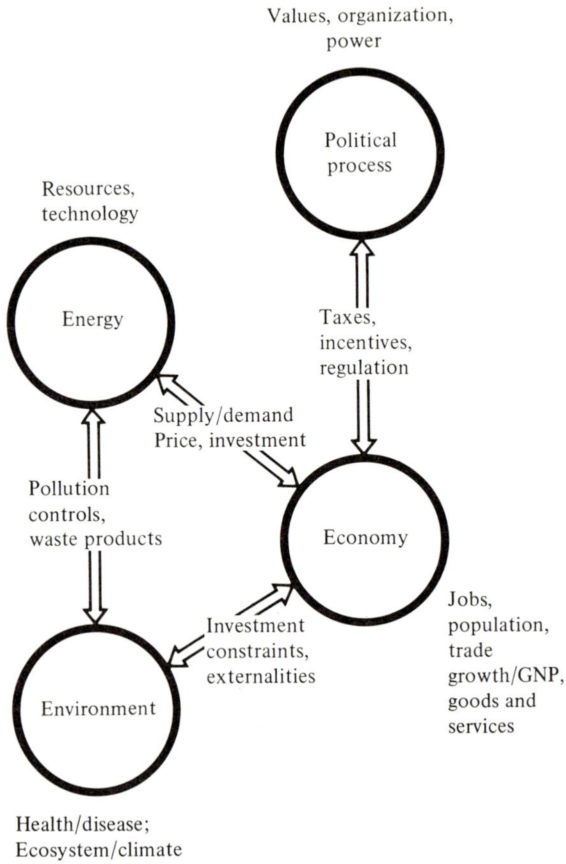

A model of the interactions of the social process of an industrialized nation portrays the important interaction of energy, economics, and the environment. Energy and economics interact directly through supply and demand, prices and capital investments. The energy marketplace is a result of the population, investment capital, governmental policies, technology, energy sources, and energy demand. The marketplace balances all these factors and yields an equilibrium of supply and demand by means of prices and controls. As the world's energy marketplace changes, a redistribution of economic wealth occurs.

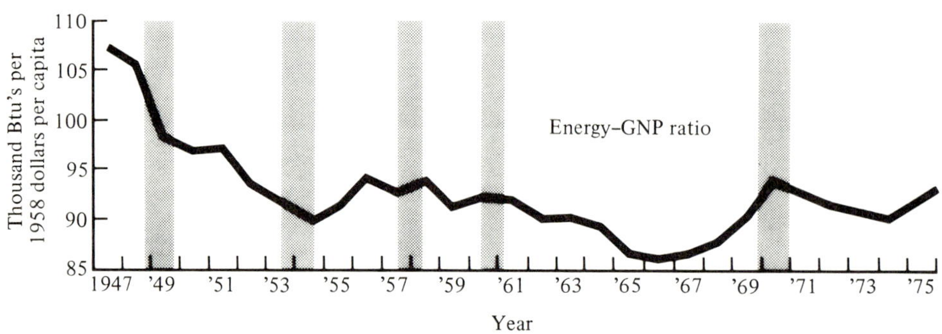

The energy-GNP ratio for the U.S. for the period 1947 to 1975. The energy used per unit of GNP rose during the period 1969–70 and again during 1973–74 due to the recessions during periods when the real GNP declined slightly.

RELATIONSHIP BETWEEN ENERGY CONSUMPTION AND GNP

Nation	b	R^2
India	1.89	0.97
Italy	1.77	0.98
Canada	1.37	0.99
Brazil	1.19	0.98
Japan	0.96	0.99
U.S.A.	0.96	0.99
W. Germany	0.66	0.95
United Kingdom	0.48	0.84

Source: P.E. de Janosi and L. E. Grayson, "Patterns of Energy Consumption and Economic Growth and Structure," *Journal of Development Studies,* Jan. 1972, pp. 241–249.

A quantitative study of the relationship of energy consumption to GNP was completed for thirty nations using the mathematical relation

$$\log E = a + b \log GNP$$

where E = energy consumption in millions of metric tons. The statistical reliability of the regression was uniformly high. However, there was a wide range in the factor b, often called the income elasticity. The value of b, and the correlation coefficient R^2 are given. Note that the income elasticity is somewhat related to the nation's stage of economic development among other factors.

THE ENERGY – GNP RATIO FOR SELECTED COUNTRIES IN 1971

Nation	*Energy per Dollar of GNP (Kilograms of Coal Equivalent per 1971 Dollars)*
USSR	3.30
Norway	2.69
Canada	2.63
United Kingdom	2.31
U.S.A.	2.24
Netherlands	1.96
W. Germany	1.65
Japan	1.62
Italy	1.53
France	1.23

ELASTICITY OF DEMAND FOR ELECTRICITY

Factor	Residential	Consumer Class Commercial	Industrial
Price	−1.3	−1.5	−1.7
Population	+0.9	+1.0	+1.1
Income per Capita	+0.3	+0.9	+0.5
Percent of Response in 1st Year	10%	11%	11%
Years for 50% of Total Response	8	7	7

The change in the demand for a commodity in response to its market price, p, is called its demand elasticity, e, in response to demand, q. So that

$$e = \frac{\Delta q/q_o}{\Delta P/P_o}$$

The value P_o and q_o are the initial values.

NET ENERGY RATIOS FOR SELECTED ENERGY PROCESSES IN THE U.S.

Energy Process	NER
Light Water Nuclear Reactor	9
Domestic Natural Gas	60
Coal Gasification	7
Oil from Oil Shale	2.8
Geothermal Dry Steam to Electricity	12.6
Electricity from Oil	0.30
Imported Oil	6

Net energy is defined as the amount of energy that remains for consumer use after the energy costs of finding, producing, upgrading, and delivering the energy has been accounted for. One measure of net energy is the net energy ratio which is defined as

$$\text{NER} = \frac{\text{delivered energy}}{\text{energy sequestered}}$$

where the energy sequestered is the energy subsidy for the process.

THE AVERAGE PRICE FOR FOSSIL FUELS IN THE U.S.

	1970	*1974*	*1978*
Petroleum to the Refinery			
Price ($/Barrel)	3.58	9.07	12.48
Price ($/$10^6$ Btu)	.60	1.50	2.70
Petroleum to Electric Utilities			
Price ($/Barrel)	3.58	11.56	13.20
Price ($/$10^6$ Btu)	.60	1.88	2.20
Coal to Electric Utilities			
Price ($/Ton)	7.70	14.50	22.88
Price ($/$10^6$ Btu)	.35	.66	1.15
Natural Gas to Utilities (Interstate)			
Price ($/1,000 ft^3)	.23	.44	1.49
Price ($/$10^6$ Btu)	.23	.44	1.49

AVERAGE RETAIL ELECTRICITY PRICES IN CENTS PER kWh IN THE U.S.

Year	*Residential*	*Industrial*	*Total*
1971	2.32	1.10	1.77
1973	2.54	1.25	1.96
1975	3.51	2.07	2.92
1978	4.48	2.79	3.72

THE AVERAGE PRICE OF NATURAL GAS IN 1971 AND 1978 IN THE U.S. FOR VARIOUS SECTORS

	Price ($/Thousand Cubic Feet)	
	1971	*1978*
Wellhead Price	.16	.90
Industrial Price	.32	1.52
Commercial Price	.76	2.02
Residential Price	1.03	2.55

ENERGY TRANSMISSION COSTS

Form of Energy	*Transmission Mode*	*Transmission Cost (¢/10^6 Btu/100 Miles)*
Natural Gas	Pipeline	1.3
Hydrogen	Pipeline	3.0
Electricity	Overhead Line-500Kv	4.2
Electricity	Overhead Line-200Kv	7.8

THE RETAIL PRICE OF GASOLINE IN THE U.S. EXPRESSED EXCLUSIVE OF TAX IN U.S. DOLLARS (Constant 1967 Dollars)

Year	*Price (Constant 1967 Dollars)*
1920	0.49
1930	0.39
1940	0.44
1950	0.37
1960	0.35
1970	0.30
1973	0.30
1975	0.40
1980 (Projected)	0.50

THE COST OF OIL IMPORTS IN THE U.S.

Year	*U.S. Oil Imports (Millions of Barrels)*	*Cost (Millions of Dollars)*	*Average Cost Barrel*
1950	199	113	$ .57
1955	321	433	1.35
1960	590	1,140	1.93
1965	833	1,711	2.05
1970	1,154	3,060	2.65
1973	2,199	7,500	3.41
1974	2,400	25,100	10.46
1978	3,000	39,000	13.06

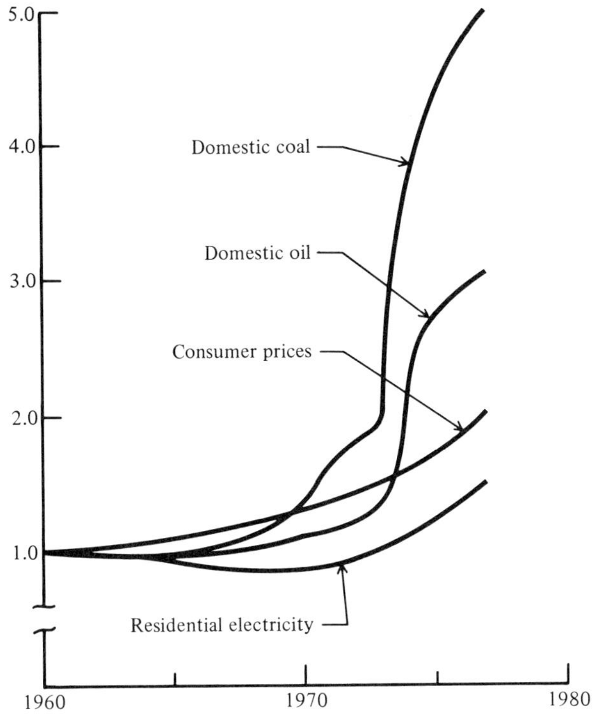

The price indices of oil, coal, and electricity compared to consumer prices. (*Source:* Statistical Abstract of the U.S., 1978.)

CONSTRUCTION COSTS OF NEW ELECTRIC POWER PLANTS FOR A PLANT LARGER THAN 800 MEGAWATTS

	Time to Construct (Months)	*Cost (Dollars per kW)*
Gas Turbine	16	$200
Oil Fired Steam	56	413
Coal Fired Steam	58	526
Nuclear	96	800

A BLOCK RATE STRUCTURE FOR GAS AND ELECTRICITY SUPPLIED TO RESIDENCES

1. *Electricity*	*Charge per Month*
Customer Charge	65¢
First 100 kWhr, per kWhr	5¢
Next 100 kWhr, per kWhr	4¢
Next 100 kWhr, per kWhr	3¢
Next 700 kWhr, per kWhr	2¢
2. *Gas*	*Charge per Month*
Customer Charge	75¢
First 25 Therms, per Therm	14¢
Next 25 Therms, per Therm	12¢
Next 25 Therms, per Therm	10¢
Over 75 Therms, per Therm	9¢

The typical rate structure for public utilities today is called the *block rate structure* and results in an average cost per energy unit (kW-hr or therm) as consumption increases. The average cost declines with increased usage due principally to increased utilization of the distribution system and the fixed cost portion of the system without proportionate increases in cost. Note that the price per unit of energy declines as consumption increases. One therm = 10^5 Btu.

A THREE PART RATE INVERTED STRUCTURE

	Charge per Month
Customer Charge	$2.00
First 300 kWhr, per kWhr	.02
Over 300 kWhr, per kWhr	.04

A three part rate structure is given in the Table. This structure is a form of what is called an *inverted rate structure* since rate per energy unit increases as consumption increases. The customer charge is used to recover the fixed costs. An inverted rate structure is useful if the aim is to discourage expansion of energy use. However, an inverted rate structure would provide a substantial economic incentive for large volume industrial users to turn to alternative sources of energy rather than pay the higher increasing rate.

THE PROFITS OF THE U.S. OIL COMPANIES IN MILLIONS OF DOLLARS

	1st Quarter 1978	*1st Quarter 1979*
Exxon	695	955
Mobil	241	437
Standard of Indiana	274	349
Standard of California	243	347
Texaco	169	307

The major oil companies are vertically integrated in that they operate at four levels: production, refinement, transportation, and marketing. Furthermore, many of these energy companies have undertaken horizontal expansion into primary energy areas, other than oil or gas, such as coal or uranium.

By 1970, the top 20 U.S. petroleum companies controlled more than 70% of the U.S. domestic production. Nevertheless, the largest share of the market held by one company, Exxon, was only 8.5% in 1975. With the 10 largest companies each holding less than 10% of the market competition is still able to thrive.

THE FIXED ASSETS OF THE U.S. OIL INDUSTRY IN 1974

Segment	*Assets (Billions of Dollars)*	*Percent of Total*
Production	$ 71	44%
Refining	32	20
Transportation of Oil	16	10
Marketing	26	16
Chemicals	11	7
Other	5	3
TOTAL	$161	100%

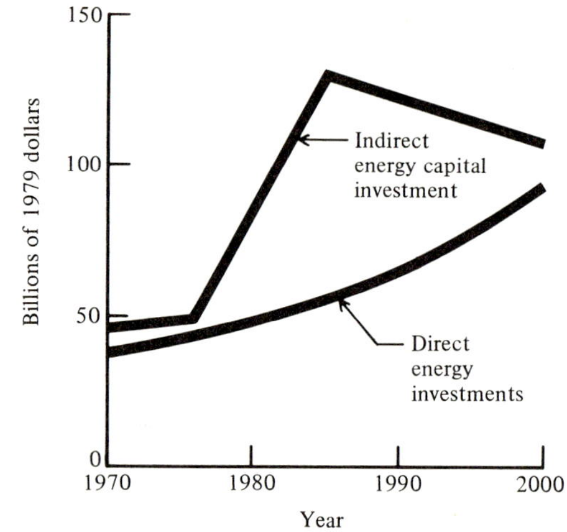

A projection of the investment capital requirements of the energy industry. The direct energy capital needs are for new utility plants, synthetic fuel plants and LNG plants, for example. The indirect energy capital requirements are investment outlays to replace existing facilities made obsolete by higher energy costs. (*Source: Business Week*, Sept. 3, 1979, pp. 175.)

RETURN ON NET WORTH IN MANUFACTURING

Year	*Total*	*Petroleum*
1971	10.8%	11.2%
1972	12.1	10.8
1973	14.9	15.6
1974	15.2	19.6
1975	12.6	13.9
1976	15.0	14.8
1977	14.9	14.0
1978	15.9	14.3

Source: Wall Street Journal, May 23, 1979, pg. 18.

COST OF ALTERNATIVE ENERGY SUPPLY

Production Costs per Barrel of Oil Equivalent (1979)	
Production Cost of Mexican Oil	$ 2.50
Production Cost of Alaskan Oil	7.50
Posted Price of OPEC Oil	18.50
LNG Imports	19.00
Low Btu Gas from Coal	20.00
Liquids from Coal or Oil Shale	24.00
High Btu Gas from Coal	25.00

20
INTERNATIONAL FACTORS

The rapid increase in demand for energy has caused great concern for the adequacy of long-range supply, not only in the U.S. but in the world as a whole. The growth in the use of energy in the world has increased at a rate of 5% per year since 1860 with a brief interruption during the depression years of 1930 to 1935. Currently, coal, oil, and natural gas provide the great predominant shares of the world's supply of energy. Wood and hydropower supply about equal shares of the world's energy supply. Petroleum is the leading source of energy in the world today, supplying about 46% of the world's energy consumed in the form of fossil fuels, hydropower, or wood. With the use of OPEC and the great dependence of many nations on imported oil and gas, the world can be viewed as one large energy marketplace.

WORLD ANNUAL CONSUMPTION OF ENERGY

Year	*Consumption ($\times 10^{15}$ Btu)*
1860	4.03
1880	9.79
1900	22.59
1920	41.89
1940	67.86
1960	122.72
1970	216.9
1976	271.0
1980 (Projected)	300.0
2000 (Projected)	522.0

PER CAPITA HOUSEHOLD ENERGY CONSUMPTION IN MAJOR COUNTRIES, 1975 (in 10^3 Kilocalories)

Country	*Energy Consumption*	*Country*	*Energy Consumption*
United States	26,304	France	11,253
West Germany	14,774	Italy	7,408
United Kingdom	13,420	Japan	7,173

PER CAPITA USE OF ENERGY IN 1970

Country	*Per Capita Use of Energy (kg of Coal Equivalent)*	*Per Capita Use of Energy (10^6 Btu)*	*Percent Increase 1965 to 1970*
USA	11,128	294	21
Canada	8,997	238	25
Britain	5,358	148	5
USSR	4,436	117	24
Japan	3,215	85	80
World Average	1,883	52	
Mexico	1,203	32	25
Saudi Arabia	827	22	126
India	189	5	8

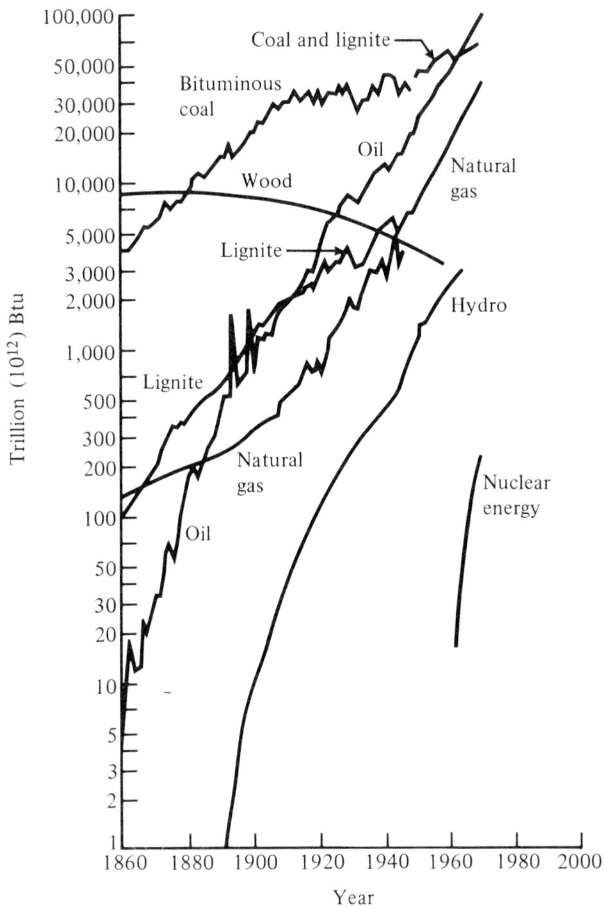

Annual world energy production of individual energy sources.

PER CAPITA USE OF ENERGY IN THE WORLD
(kg of Coal Equivalent per Person)

1970	1881
1972	1997
1974	2017
1976	2069

Source: U.N. Statistical Yearbook 1977, Section 12.

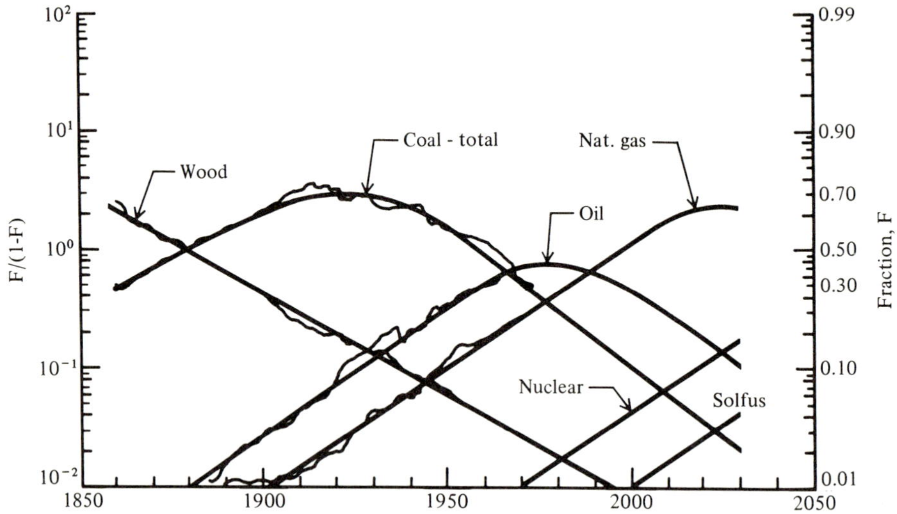

Market penetration of primary energy sources in the world. F is the market fraction taken by a given source at a certain time. Wriggling lines report statistical data. Smooth lines are the logistic fit of these data. Actual values of fractions are reported on the right side of the graph for seven points. In the case of nuclear, the penetration is still too low to permit fitting of the equation. Consequently the rate of penetration is assumed to be equal to that of other primary energies. SOLar-FUSion are purely hypothetical and are shown only after the year 2000. Note that natural gas is the primary supply after 1990. (*Source:* C. Marchetti, "Why Hydrogen", *Int. J. Hydrogen Energy,* Vol. 4, 1979, pp. 1–5.)

ENERGY CONSUMPTION IN 1976

Region	*Total Energy Consumed (Millions Metric Tons of Coal Equiv.)*	*Per Capita Energy (kg Coal Equiv./Person)*
Africa	164	397
Central America	185	1,265
South America	158	868
North America	2,716	11,395
Middle East	8,144	1,169
Asia	704	557
W. Europe	1,573	4,289
Oceania	106	4,953
Centrally Planned Economies*	2,570	2,030
World	8,318	2,069

*E. Europe, P. R. China, USSR

ELECTRIC ENERGY PRODUCTION IN BILLIONS OF kWh IN 1976

Region or Country	*Production*
Africa	139
North America	2,506
South America	184
Asia	961
W. Europe	1,913
USSR	1,111
World	6,917

Source: U.N. Statistical Yearbook, 1977, section 12.

WORLD ENERGY SOURCES IN 1976

Source	*Supply (Millions of Metric Tons of Coal Equivalent)*	*Percent of Total*
Coal and Lignite	2,702	29.5
Oil	4,359	47.6
Natural Gas	1,662	18.1
Nuclear/Hydro	228	2.5
Wood	210	2.3
TOTAL	9,161	100%

ANNUAL IMPORTS OF PETROLEUM IN 1976

Country	*Cost of Oil Imports (Billion Dollars U.S.)*
U.S.	34.2
France	13.3
W. Germany	14.0
Britain	10.0
Japan	23.3

A COMPARISON OF ENERGY CONSUMPTION PATTERNS FOR FIVE COUNTRIES IN 1972

	Sweden	*Britain*	*U.S.A.*	*West Germany*	*Japan*
Population (Millions)	8.1	55.8	208.8	61.7	107.0
Gross National Product ($1 Million U.S.)	43.8	157.0	1,159.3	174.2	251.4
Energy Consumption (10^{15} Btu)	1.3	8.4	67.3	9.2	9.6
Energy Consumption per Capita (10^6 Btu)	152	143	307	143	86
Energy Consumption per Million of Dollars of GNP (10^7 Btu per $\$10^6$)	3.0	5.4	5.8	5.3	3.8

PETROLEUM IN WESTERN EUROPE* (EEC) IN 1975

Annual Production	0.18 Billion Barrels
Annual Consumption	4.97
Annual Imports	4.79
Total Proven Reserves	20 Billion Barrels

*European Economic Community nations are: Belgium, Denmark, France, Great Britain, Ireland, Italy, Luxemburg, Netherlands, and West Germany.

Today Europe's imported energy is in the form of oil and represents approximately 60 percent of the world trade in oil. The EEC nations were dependent upon imports for 96 percent of their petroleum in 1975. The EEC is dependent upon the importation of oil for 60 percent of their total energy needs, whereas they imported only 31 percent of their total energy consumed in 1960. As western Europe's energy use has grown at an annual rate of 5 percent over the past two decades, the dependence upon imported oil has grown proportionately. Petroleum supplies 60 percent of the total energy consumed in western Europe.

TOTAL OIL IMPORTS FOR THREE NATIONS AS A PERCENT OF THE GROSS NATIONAL PRODUCT

Year	*U.S.*	*W. Germany*	*Japan*
1970	0.3%	1.3%	1.1%
1972	0.4	1.3	1.3
1974	1.7	3.3	4.2
1976	1.8	3.1	3.8
1977	2.2	2.9	3.4

W. Germany reduced its oil consumption from 1974 to 1977 while Japan and the U.S. increased their annual consumption.

THE ENERGY SOURCES IN WESTERN EUROPE

Source	*Percent of Total*
Coal	29
Oil	60
Natural Gas	7
Hydroelectric	3
Nuclear	1
TOTAL	100

Natural gas, as well as petroleum, is used in western Europe and supplies 7 percent of the total energy consumed. About 10 percent of this natural gas is imported from Algeria and the USSR (via Czechoslovakia). The primary source of natural gas in the EEC is the Netherlands.

The total proved reserves of coal in western Europe are estimated to be 3.5×10^{18} Btu (1.5×10^{11} tons), which is sufficient to supply Europe for many decades at the present rate of consumption.

U.S. FOREIGN NET TRADE FOR 1950 TO 1978

Year	*Exported Coal (Million Tons)*	*Imported Oil and Refined Products (Million Barrels)*	*Imported Natural Gas (10^9 Cubic Feet)*
1950	29.0	198.9	(25.7)*
1960	37.7	590.2	144.3
1970	72.4	1,153.5	751.0
1973	52.8	2,158.0	950.0
1978	33.3	2,880.0	940.0

*25.7×10^9 cu ft of gas were exported in 1950.

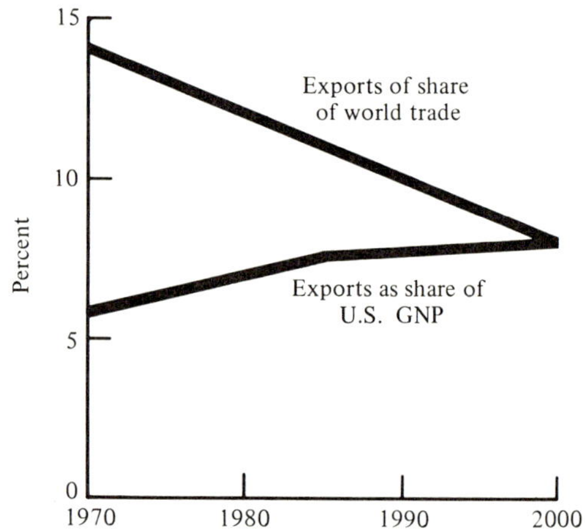

U.S. exports will provide a larger share of the U.S. GNP but will decline as a share of world trade. The U.S. exports agricultural and technological products and imports energy and minerals. (*Source: Business Week,* Sept. 3, 1979, pg. 177.)

ECONOMIC ACTIVITY AND ENERGY IN 1972 FOR SELECTED COUNTRIES

Country	*Energy Consumption (Lbs. of Coal Equiv./Dollar)*	*Automobile Use (Autos 10^3 Passenger-km)*
U.S.	4.64	7.17
New Zealand	2.54	1.88
Norway	2.71	0.34
Japan	2.60	0.04

ENERGY CONSUMPTION IN GREAT BRITAIN IN 1973 AND PROJECTED FOR 1985

Source	*1973* Consumption (Million Tons of Coal Equivalent)	Percent of Total	*1985* Consumption (Million Tons of Coal Equivalent)	Percent of Total
Coal	131.3	38	135	31
Petroleum	159.4	46	175	41
Natural Gas	43.5	12	80	19
Nuclear	9.9	3	40	9
Hydroelectric	2.0	1	2	–
TOTAL	346.1	100	432	100

Several nations such as Sweden and Switzerland may have already passed the U.S. with respect to GNP per capita. The reason is that these countries do not have to divert so much capital away from investment in growth-promoting activities towards dealing with the consequences of excessive energy use and low energy prices as in the U.S. Nations with extensive hydroelectric power available, such as Norway, have developed a system of electric street cars and electric trains. This form of transportation mitigates against urban sprawl and wasteful use of energy by automobiles.

NUCLEAR POWER REACTORS OF SELECTED NATIONS IN 1977

Country	Operating	Under Construction	Ordered	Planned	Total
Canada	7	10	4	5	26
France	10	17	12	8	47
W. Germany	7	12	8	4	31
Japan	10	14	0	5	29
USSR	19	8	0	10	37
U.S.	72	10	12	18	112
United Kingdom	29	10	0	7	46
50 Countries	190	130	75	200	595

CRUDE OIL PRODUCTION OF THREE COUNTRIES WITH RECENT INCREASES IN PRODUCTION

Nation	1970	1976	1980	1985 (Projected)
USSR	7.1	9.8	11.8	12.0
China (PRC)	0.6	1.7	2.6	4.0
Mexico	0.4	0.9	2.2	4.0

Production in millions of barrels per day.

THE CHANGE IN THE PER CAPITA USE OF OIL IN SELECTED NATIONS FOR THE PERIOD 1973 TO 1978

Nation	Percent Change in Oil Consumed per Capita
U.S.	+17.1%
Canada	+10.9%
Japan	+ 6.2%
W. Germany	− 2.3%
France	− 6.2%
Britain	−16.9%

Source: U.S. News and World Report, July 16, 1979, pg. 22.

COMPARATIVE ECONOMIC INDICATORS FOR DEVELOPED NATIONS FOR 1979

	U.S.	Japan	W. Germany	France	Britain	Italy	Canada
GNP Growth (%) 1st Quarter 1979	.8	7.4	−6.4	3.0	−10.0	3.6	7.2
Inflation (%) Jan.-April 1979	13.2	7.8	7.4	10.8	13.8	17.4	11.7
1978 Net Oil Imports ($ Billion)	39.2	24.3	13.2	10.7	4.0	9.1	1.1
1978 Percent of Oil Imported	47	92	97	97	44	97	14

GASOLINE PRICES AROUND THE WORLD IN 1979 IN DOLLARS PER U.S. GALLON

Nation	*Price ($/U.S. Gallon)*
U.S.	$0.95
Britain	1.80
Japan	1.98
France	2.45
Greece	2.93

GASOLINE TAXES, 1978

Country	*Per Gallon*	*Increase 1973–78*
Belgium	$1.18	22%
France	1.10	62
Germany	1.02	6
Italy	1.54	161
Netherlands	1.12	19
Spain	.35	30
Sweden	.69	−4
U.K.	.75	108
U.S.	.12	None

Source: Statistical Abstract, 1978, p. 655.

COMPARATIVE ENERGY/OUTPUT RATIOS FOR 1976

Country	*Gross Domestic Product Per Capita (Dollars)*	*Energy Consumption Per Capita (Tons of Oil Equiv.)*	*Energy/GDP (Tons of Oil Equiv. $ Million GDP)*	*Index (U.S. = 100)*
U.S.	5,960	8.3	1,390	100
France	4,740	3.5	750	54
W. Germany	4,350	4.5	1,020	73
Sweden	5,460	6.1	1,190	82

Source: R. Stobaugh and D. Yergin, *Energy Future,* Random House, 1979, pg. 143.

GLOBAL DISTRIBUTION OF ENERGY USE IN 1972 AND RECENT RATES OF CHANGE

	Less Developed Countries	*All Developed Countries*	
		Other Developed Countries	*U.S.A.*
Population (Billions)	2.5	1.0	0.2
Per Capita Energy Use (10^6 Btu/Person)	13.7	100	307
Total Energy Use (10^{15} Btu)	36.5	103.8	67.3
Percent of Total Energy Use	18%	50%	32%
Recent Annual Growth of:			
Population	2.5%	1.0%	0.7%
Total Energy Use	5.5%	5.0%	4.0%
Per Capita Energy Use	3.0%	4.0%	4.0%

One way of examining the possibility for a transition of energy usage throughout the world is to divide the world into two groups of nations. We will denote the relatively well-off nations as wealthy and the poorer, less well-off or less developed countries (LDC) as the other group. The 2.5 billion people living in countries with per capita gross national product less than \$650 1972 U.S. dollars will be denoted as the less developed countries (LDC). The remaining 1.2 billion people living in nations with a per capita GNP greater than \$650 will be denoted as developed countries. The global distribution of energy use for this oversimplified division is given in the table where the developed nations are divided into the U.S. and other nations. *Source:* J. P. Holdren, "Energy and Prosperity," *Bulletin of Atomic Scientists,* Jan. 1975, pp. 26-28.

AN OPTIMISTIC SCENARIO FOR THE GROWTH OF ENERGY USE FOR THE PERIOD 1975 TO 2005

	Developed Countries		*Less Developed Countries*	
	Population	*Per Capita Energy Use*	*Population*	*Per Capita Energy Use*
Growth Rates				
1975-1985	0.8%	3.0%	2.0%	2.5%
1985-1995	0.3%	1.0%	1.5%	3.5%
1995-2005	0.0%	0.2%	1.0%	4.5%
Values at Specified Times				
(Billions)		(10^6 Btu)	(Billions)	(10^6 Btu)
1975	1.25	150	2.75	15
1985	1.35	192	3.41	20
2000	1.44	212	4.31	36

In order to permit an acceleration of energy use by the less developed nations, a slowdown in the growth of use by industrialized nations would be required. However, this is a possibility since population growth in the developed nations has slowed down to less than 1% per year. Maintaining a constant per capita use of energy in the developed nations would enable the less developed nations to significantly expand their use of energy. One optimistic scenario for redistribution of the growth of energy use calls for a reduced rate of growth of population of the LDC over the next 3 decades while the per capita use of energy increased. Also, this projection assumes a zero rate of population growth of developed nations by the year 2000 and a leveling off in the rate of growth of per capita energy use. Then, with this transition in the rates of energy growth, the per capita energy use would increase by 41% in the developed nations over the 30-year period while increasing by 140% in the less developed nations. Using this projection the energy consumption in the year 2000 would be 305×10^{15} Btu and 155×10^{15} Btu in the developed and less developed countries respectively. Then, the total world consumption in 2000 would be 460×10^{15} Btu compared with 230×10^{15} Btu in 1975.

THE WORLD'S HYDROELECTRIC RESOURCE ENERGY POTENTIAL

	Annual Energy Output (10^9 kWh)
China	1,320
USSR	1,095
USA	701
Zaire	660
Canada	535
Brazil	519
India	205
WORLD TOTAL	9,802

Perhaps one of the best areas of potential development of energy for less developed nations lies in the construction of hydroelectric projects. Asia and Africa hold vast resources of potentially developable hydropower projects. However, hydroelectric projects are more capital-intensive than many other kinds of power plant, and require long construction times. Therefore, less developed nations often are able only slowly to develop their hydroelectric potential. If one-quarter of the world's hydroelectric resource could be exploited by the year 2000, then approximately 2400×10^9 kWh (8.2 quads) could be obtained annually. This, unfortunately, might be only 3 or 4 percent of the world's total energy consumption.

ESTIMATED WORLD RESERVE OF FOSSIL FUELS EXPRESSED IN QUADRILLIONS OF Btu's (10^{15} Btu)

Coal	15,000 Q
Petroleum	3,500 Q
Natural Gas	2,000 Q
Uranium Ore	800 Q
TOTAL	21,300 Q

The world consumption of energy in 1979 was about 290 quads and the estimated known reserves is 21,300 quads. Thus, the reserve to production ratio is approximately 75.

ESTIMATED WORLD RESERVES AND RESOURCES OF FOSSIL FUELS

	Known Recoverable Reserves	*Total Resource*
Coal (10^9 Tons)	665	10,000
Petroleum (10^9 Barrels)	556	2,000
Natural Gas (10^{12} Cubic Feet)	2,230	13,000
Shale Oil (10^9 Barrels)	—	2,000
Uranium Ore (10^3 Tons)	1,085	4,000

The term *resources* includes reserves, but also encompasses deposits already identified but not presently recoverable as well as undiscovered deposits which may or may not be recoverable. (*Source:* V. E. McKelvy, "World Energy Reserves and Resources", *Public Utilities Fortnightly,* Sept. 25, 1975, pp. 27–33.)

THE WORLD'S PROVEN COAL RESERVES (IN BILLIONS OF TONS)

U.S.S.R. and other Communist Nations	305
United States	216
Western Europe	72
Far East and Oceania	46
Africa	17
Other	9
TOTAL	665

The known reserves of fuels are not evenly distributed throughout the world. The Middle East and Africa hold about 69% of the world's proven reserves of fuels. Note that the U.S. holds about 32% of the world's proven coal reserves.

ENERGY RESOURCES OF THE THREE NORTH AMERICAN NATIONS

Country	*Population (Millions)*	*GNP Per Capita*	*Proven Oil Reserves (Billions of Barrels)*	*Oil Consumption (Millions of Barrels/Day)*	*Gas Reserves (Trillion Cu. Feet)*	*Annual Gas Consumption (Trillion Cu. Feet)*
Canada	23.7	$8,450	6.9	1.7	66.4	1.5
U.S.	220	8,640	27.8	18.8	200.3	19.4
Mexico	67.9	1,374	26.0	0.9	58.9	0.025

Source: H. E. Meyer, "Why a North American Common Market Won't Work," *Fortune,* Sept. 10, 1979, pp. 118–122.

CANADIAN RESOURCES

	Resource	*Proven Reserve*
Petroleum (Billions of Barrels)	121	6.9
Natural Gas (10^{12} Cubic Feet)	780	66.4
Alberta's Tar Sands (Billions of Barrels)	300	

Of the estimated Canadian resources of 121 billion barrels of petroleum, only seven billion barrels are proven reserves. Canada also holds an estimated 66×10^{12} cubic feet of proven reserves of natural gas. The western province of Alberta produces 85 percent of Canada's oil and 88 percent of its natural gas. In addition, the untapped tar sand resource lies in Alberta.

Hydroelectric power is another important part of Canada's energy resource. The total installed hydroelectric capacity exceeds 34 gigawatts, representing 63 percent of the nation's total electric-power capacity. Hydroelectricity provides about 23 percent of Canada's primary energy supply. This is in contrast to the U.S., where hydroelectric energy supplies less than 2 percent of the total energy consumed in the nation. Over the next decade, Canada is planning to further develop hydroelectric projects and explore storage schemes. In 1975, Canada's imports of oil were 950,000 barrels per day where exports were 700,000 barrels per day. An abundance of natural resources makes Canada almost self-sufficient in the supply of energy. Yet the country's vast reserves of coal, oil, and tar sands and its large potential for hydroelectric development do not insulate the nation from the possibility that it will experience an energy shortage.

RUSSIAN FOSSIL FUEL RESERVES AND RESOURCES

	Proved Reserve	*Estimated Resource*
Coal (10^{12} Metric Tons)	2.2	7.8
Petroleum (Billions of Barrels)	47.5	Not Available
Natural Gas (10^{12} Cubic Meters)	24	100

COMPARATIVE INDICES IN 1979

Index	*U.S.*	*USSR*	*China (PRC)*
Population (Million)	220	262	1,000
GNP (Billion Dollars)	2,100	1,200	444
Annual Oil Production (Million Tons)	474	629	110
Annual Coal Production (Million Tons)	654	796	651
Annual Energy Consumption per Capita ($\times 10^6$ Btu/Person	310	130	20

Source: Newsweek, Feb. 5, 1979, pg. 43.

21
ENERGY POLICY

The national security, financial stability, and standard of living of a nation are intertwined with its energy consumption. Recent international events have brought many nations sharply to the knowledge that domestic inflation and related economic factors, adjustments in life style, and national security can be challenged and impacted by the reduced availability of energy supplies. Thus, many nations are finding it necessary to inquire what energy and environmental policies are necessary to balance supply and energy consistent with acceptable economic, social, and environmental goals.

Policy issues are numerous for the U.S. and include those listed in the next section. Action on many of these policy issues will be required over the next decade in order to avoid significant unforeseen impacts on the national economy and standard of living. It is critical to reconcile environmental policies with energy policies so they are coordinated, not counterproductive. Also, decisions on the balance of foreign and domestic oil to be used over the next decade is an important issue. The control and regulation of energy prices is an issue of large magnitude. Should large energy companies be allowed to grow and possibly dominate the energy marketplace?

ENERGY POLICY ISSUES FOR THE U.S.

1. Supply and Demand of Energy
   a) Should the U.S. adopt a policy of self-sufficiency in energy resource development and production? Or should the U.S. pursue a policy of maximum importation of energy sources in order to conserve domestic supplies?
   b) How dependent can the U.S. become on imported petroleum?

Should the nation pursue a policy designed to increase the development of its urban, onshore, and offshore oil reserves?

c) Should the U.S. seek ways to negotiate contracts that guarantee delivery of imported natural gas?

d) Should tax and fiscal policies be designed to provide increased incentives for the development of potential energy reserves? To increase oil refining capacities?

e) Can synthetic fuels close the energy gap? What are the prospects for solar power, Geothermal, Magneto hydro-dynamics, Fusion or other alternate sources?

f) What are the air quality implications of increased use of low sulfur oil, high sulfur oil, and coal?

g) Should the nation adopt policies to alter existing building codes that would reduce energy demand? How effective would it be?

h) How can the energy demand of the transportation market be lowered?

2. Policy Implementation

a) Are national security, clean energy and a low cost national energy base compatible?

b) Should the price of energy be allowed to be set by a free-market system or an alternate system?

c) What are the implications of decisions for siting refineries and associated industries in connection with deep-water terminals to receive imported energy supplies?

d) Should there be established a "one-step" approval procedure for energy-related projects (i.e., power plant siting, urban and offshore oil drilling, geothermal development, etc.)?

e) Should a coordinated federal-state energy research and development effort be developed?

PRIMARY OBJECTIVES OF THE U.S. ENERGY POLICY

1. The development of an adequate supply of energy at reasonable prices to permit the nation to enjoy a good standard of living.
2. The achievement of relative self-sufficiency of energy supply.
3. The maintenance of a safe and healthy environment.
4. The attainment of maximum efficiency in the production, distribution, and utilization of all forms of energy.

HISTORY OF THE U.S. FEDERAL GOVERNMENT'S INVOLVEMENT IN ENERGY POLICY

1. The Organic Act of 1849 established the Department of Interior to manage the public lands and the resources of these lands.
2. In 1866, the Mining Act declared these public lands to be "free and open" to mining.
3. The U.S. Geological Survey was created in 1879 to classify and examine the geological structure and the mineral resources of the nation.
4. Under the Coal Land Act of 1873, lands known to contain valuable coal deposits were sold in small tracts for mining purposes.
5. The U.S. Department of Commerce was founded in 1903 with an objective, in part, to foster, promote, and develop domestic commerce and mining of the U.S.
6. The Bureau of Mines Act of 1910 created the Bureau to conduct investigations concerning mining and mineral substances.
7. The Federal Water Power Act of 1920 established the Federal Power Commission, later reorganized as an independent agency in 1930. In 1935, the FPC was given authority over interstate transmission of electric energy and the investigatory and advisory functions extended to the whole electric industry.
8. The 1938 Natural Gas Act extended the FPC controls to include interstate transmission and sales of natural gas.
9. The Atomic Energy Act of 1946 established a civilian Atomic Energy Commission (AEC) with control over all policy aspects of atomic energy development.
10. The Atomic Energy Act of 1954 gave the AEC authority over the development, licensing, and regulation of private power reactors.
11. In 1970, the President's Reorganization Plan created the Environmental Protection Agency (EPA).
12. In 1974 a bill was passed by Congress that split the Atomic Energy Commission into two agencies—an Energy Research and Development Administration (ERDA) and a new regulatory body, the Nuclear Regulatory Commission (NRC). The objectives of this organization were to: (1) reduce the inflexibility and lack of coordination among federal agencies and regulators in the energy field, and (2) improve the completeness, reliability, and accuracy of energy data.
13. A new cabinet-level Department of Energy incorporated ERDA and the energy activities of other federal agencies commenced operation in October 1977 with a budget of 11 billion dollars.

A SUMMARY OF THE ERDA PLAN GOALS

Energy Source	*U.S. Energy Conversion Goals (10^{15} Btu)* 1985	2000
Coal Gasification and Liquification	0	9
Direct Coal Utilization	6	9
Oil Shale	2.0	4
Enhanced Oil and Gas Recovery	6	9

ERDA, now part of the Department of Energy, released a national plan for energy in 1975 which called for major changes in U.S. energy policy. In this plan, the pursuit of commercially feasible solar energy systems was elevated to the highest priority. This was a change from a former priority commitment to nuclear power as the sole source of new energy alternatives.

THE U.S. BUDGET FOR ENERGY RESEARCH AND DEVELOPMENT*
(Millions of Dollars)

Program Area	*Fiscal Year* 1973	1974	1975	1976	1978
Conservation	32.2	65.0	71	91	100
Oil, Gas and Shale	18.7	19.1	43	44	105
Coal	85.1	164.4	418	442	659
Environmental Control	38.4	65.5	185	199	238
Nuclear Fission	406.5	530.5	625	709	1,113
Nuclear Fusion	74.8	101.1	264	304	332
Solar	4.0	13.8	89	116	372
Geothermal	4.4	10.9	32	50	108
Systems Studies	7.2	17.3	50	60	70
Other	0.9	11.5	203	398	200
TOTAL	672.2	999.1	1,980	2,413	3,297

*Some line items are not fully comparable due to changes in categories for reporting.

U.S. DEPARTMENT OF ENERGY BUDGET

	Budget Authority ($ in Millions)		
	1978	*1979*	*1980*
Energy Technology	$ 3,373	$ 3,625	$3,583
Basic Sciences	406	431	474
Conservation	538	671	555
Regulation and Information	182	276	323
Defense Activities	2,562	2,685	3,022
Government-Owned Operations	738	248	149
Policy and Management	241	258	308
Less: Supplemental and Other Adjustments	−122	−403	—
SUBTOTAL	$ 7,918	$ 7,791	$8,415
Strategic Petroleum Reserve*	3,180	3,008	8
TOTAL DOE FUNDING	$11,098	$10,799	$8,423

*Because much of the $3 billion provided for the Strategic Petroleum Reserve (SPR) in FY 1979 remained available, no additional SPR funding was requested for 1980 except $8 million for operational expense.

THE U.S. ENERGY PLAN PROPOSED BY PRESIDENT CARTER IN APRIL AND JULY, 1979

1. Phase out price controls on U.S. crude oil by September 30, 1981.
2. Levy a 50 percent excess-profits tax (windfall tax) on revenue from decontrolled oil.
3. Divide the estimated 10 billion dollars in excess-profits tax receipts among energy research, mass transit and assistance for low-income families.
4. Limit thermostat settings to 65 degrees in winter and 80 degrees in summer for nonresidential buildings.
5. Phase out free or reduced-rate parking for federal workers around the country.
6. Allow tax credits for purchase of wood-burning stoves, passive solar-energy systems and for production of shale oil.
7. Set annual limits on oil imports beginning with a ceiling of 8.2 million barrels per day for 1979. Cut the import of oil to 4 million barrels per day by 1990.
8. Establish a government chartered energy corporation to develop a synthetic fuel industry. The goal is 2.5 million barrels per day of oil from shale or coal.
9. Establish a standby plan for gasoline rationing.

10. Spend 10 billion dollars to improve bus and rail systems and 6.5 billion dollars to upgrade the gasoline efficiency of automobiles.
11. A government "solar bank" will be set up to subsidize loans for installation of solar-energy systems in both homes and business. Tax credits will also be provided.

THE PLANNED COST OF PRESIDENT CARTER'S 1979 ENERGY PLAN

Cost for the Period 1980–1990 (billions of dollars)	
Energy Security Corporation to Develop Alternative Fuels	$ 88.0
Aid to the Poor to Help with Higher Fuel Bills	24.0
Mass Transit—To Buy and Improve Buses and Subways	16.5
To Finance a Switch by Utilities from Oil to Other Fuels	5.0
To Stimulate Solar Power, A Solar Bank, and Tax Credits	3.5
Conservation to Save Energy in Buildings through Loans and Grants	2.0
Tax Credits for Specific Energy Sources	2.0
TOTAL	$141.0

GLOSSARY

Alternating current (ac): An electric current that reverses its direction of flow periodically (see frequency) as contrasted to direct current (dc).

Anthracite: A hard, black, lustrous coal that burns efficiently and is therefore valued for its heating quality.

Atomic Energy Commission (AEC): A five-member commission established after World War II to supervise and promote use of nuclear energy. The commission was abolished in 1975, and its functions were transferred to the Energy Research and Development Administration (ERDA) and the Nuclear Regulatory Commission (NRC).

Barrel: A liquid-volume measure equal to 42 U.S. gallons, commonly used in expressing quantities of petroleum and petroleum products (bbl).

Baseload: The minimum load of a utility (electric or gas) over a given period of time. Units in baseload service are operated in full capacity all the time.

Binary cycle: An energy-recovery system which results in heat exchange between two separate fluid-circulation systems. The purpose of a binary cycle is to obtain higher efficiencies from the energy source.

Bioconversion: A general term describing the conversion of one form of energy into another by plants or microorganisms. Synthesis of organic compounds from carbon dioxide by plants is bioconversion of solar energy into stored chemical energy. Similarly, digestion of solid wastes or sewage sludge by microorganisms to form methane is bioconversion of one form of stored chemical energy into another, more useful form.

Bituminous coal: Soft coal; coal that is high in carbonaceous and volatile matter. When volatile matter is removed from bituminous coal by heating in the absence of air, the coal becomes coke.

Breeder: A nuclear reactor that produces more fuel than it consumes. Breeding is possible because of two facts of nuclear physics: (1) Fission of atomic nuclei produces on the average more than two neutrons for each nucleus undergoing reaction. In simplified terms, then, one neutron can be used to sustain the fission chain reaction and the excess neutrons can be used to create more fuel. (2) Some nonfissionable nuclei can be converted into fissionable nuclei by capture of a neutron of proper energy. Nonfissionable uranium-238, for example, can thus be bred into fissionable plutonium-239 upon irradiation with high-speed neutrons.

British thermal unit (Btu): The quantity of heat necessary to raise the temperature of one pound of water one degree Fahrenheit. One Btu equals 252 calories, 778 foot-pounds, 1055 joules, and 0.293 watt-hours.

Calorie: Originally, the amount of heat energy required to raise the temperature of 1 gram of water 1 degree Centigrade. Because this quantity varies with the temperature of the water, the calorie has been redefined in terms of other energy units. One calorie is equal to 4.2 joules. (When capitalized, Calorie means 1000 calories (or 1 kcal). Therefore, one Calorie is equal to 4184 joules.

Central station power: Production of power—usually electrical—in large quantities at a generation plant as opposed to production at the point of consumption.

Coal: A solid, combustible organic material formed by the decomposition of vegetable material without free access to air. Chemically, coal is composed chiefly of condensed aromatic ring structures of high molecular weight. It thus has a higher ratio of carbon to hydrogen content than does petroleum.

Coal gasification: The conversion of coal to a gas suitable for use as a fuel.

Coal slurry pipelines: A pipeline which transports coal in pulverized form suspended in water.

Conduction: The process by which energy is transferred directly from molecule to molecule. It is the way in which electricity travels through a wire or heat moves from a warm body to a cool one when the two bodies are placed in contact.

Continental shelf: The submerged shelf of land sloping gradually from the exposed edge of a continent. It is usually defined as those areas where the water is less than 200 meters (600 feet) deep.

Controlled thermonuclear reactor (CTR): Controlled fusion, that is, fusion produced under research conditions, or for production of useful power.

Convection: The transfer of heat by the circulation of a liquid or gas.

Core: The central part of a nuclear reactor which contains the nuclear fuel.

Critical mass: The minimum amount and arrangement of a fissionable material, such as uranium-235 or plutonium-239, that is required to sustain fission in a nuclear reactor.

Crude oil: Petroleum liquids as they come from the ground. Also called simply "crude."

Curie: The unit of radioactivity. One curie is 3.7×10^{10} disintegrations per second. One gram of radium has 1 curie of radioactivity.

Decay, radioactive: The process whereby atoms of radioactive substances experience transformation into atoms of other elements with attendant emission of penetrating radiations (gamma ray) and some nuclear particles. Each radioactive substance has a unique decay rate, which may range from a fraction of a second to hundreds of years or more.

Depletion allowance: A tax allowance extended to the owner of exhaustible resources based on an estimate of the permanent reduction in value caused by the removal of the resource.

Deuterium: A nonradioactive isotope of hydrogen whose nucleus contains one neutron and one proton and is therefore about twice as heavy as normal hydrogen. Often referred to as heavy hydrogen.

Diesel oil: The oil fraction left after petroleum and kerosene have been distilled from crude oil.

Direct current (dc): An electric current, such as that produced by a battery, in which the electrical potential does not change its sign, and the voltage is often invariant with time. In a direct current, therefore, energy is carried by a continuous, unidirectional flow of electrons through a conductor.

Electric energy: The energy associated with electric charges and their movements. Measured in watt-hours or kilowatt-hours. One watt-hour equals 860 calories.

Electron: An elementary particle with a rest mass of 9.1×10^{-28} grams, bearing a negative electric charge. Electrons orbit the atomic nucleus; their transfer or rearrangement between atoms underlies all chemical reactions. Either negative or positive electrons (sometimes called positrons) may be emitted from atomic nuclei during nuclear reactions: they are then called beta particles.

Energy: The capacity to do work. A quantity which is conserved, although it may be exchanged among bodies and transformed from one form to another, converted between heat and work, or interconverted with mass.

Energy conversion: The transformation of energy from one form to another.

Enrichment: The process of increasing the concentration of fissionable uranium-235 in uranium from the naturally occurring level of about 0.7 percent to the concentration required to sustain fission in a nuclear reactor, generally about three percent. The principal method of enrichment is gaseous diffusion, but gaseous centrifugation is also receiving much attention, particularly abroad.

Entropy: The change in entropy, ΔS, of any system is equal to the increment of heat added divided by its absolute temperature, so that:

$$dS = \frac{dQ}{T} \frac{\text{joules}}{\text{kelvin}} .$$

The total entropy of the universe is continually increasing.

Environmental impact statements: The analytical statements that balance costs and benefits of a federal decision. Required by the National Environmental Policy Act (NEPA), section (102(2) (c).

Exponential growth: Type of growth illustrated by the compound interest law. A useful characteristic of an exponential growth rate is that the "doubling time," or the length of time required for the growing thing to double in size, is constant.

Fast breeder reactor: A nuclear reactor that operates with neutrons at the fast speed of their initial emission from the fission process, and that produces more fissionable material than it consumes.

Fast neutron: High energy neutron. Fast neutrons are utilized in the fast breeder actor both to produce nuclear fissions and to transform fertile material (e.g., ^{238}U) into fissionable nuclear fuel.

Fast reactor: A nuclear reactor in which the fission chain reaction is sustained primarily by fast neutrons. Fast reactors contain no moderator and inherently require enriched fuel. They are of interest because of favorable neutron economy which makes them suitable for breeding.

Feedstock: Fossil fuels used for their chemical properties, rather than their value as fuel, e.g., oil used to produce plastics and synthetic fabrics.

Fertile material: A material, not itself readily fissionable by thermal neutrons, which can be converted into a fissionable material by irradiation in a nuclear reactor. The two basic fertile materials are uranium-238 and thorium-232.

Fission: The splitting of a heavy nucleus into two approximately equal parts, accompanied by the release of a relatively large amount of energy and generally one or more neutrons. Fission can occur spontaneously, but usually is caused by nuclear absorption of neutrons or other particles.

Fissile materials: Any material fissionable by slow neutrons. The three basic ones are uranium-235, plutonium-239, and uranium-233.

Fly ash: The fine, solid particles of noncombustible material residue carried from a bed of solid fuel by the gaseous products of combustion.

Fossil fuel: Any naturally occurring fuel of an organic nature, such as coal, oil shale, natural gas, or crude oil. Fossil fuels are organically formed from living matter.

Fuel: A substance used to produce heat energy, chemical energy by combustion or nuclear energy by nuclear fission.

Fuel cell: A device for converting the energy released in a chemical reaction directly into electrical energy.

Fuel cycle: The series of steps involved in supplying fuel for nuclear power reactors. It includes mining, refining of uranium, fabrication of fuel elements, their use in a nuclear reactor, chemical processing to recover remaining fissionable material, reenrichment of the fuel, refabrication into new fuel elements, and waste storage. Fuel cycle is sometimes used to refer to a similar series of steps for fossil fuels.

Fusion: The combining of atomic nuclei of very light elements by collision at high speed to form new and heavier elements, resulting in the release of energy.

Gallon: A unit of measure. A U.S. gallon contains 231 cubic inches, 0.133 cubic feet, or 3.785 liters. It is 0.83 times the imperial gallon.

Gas-cooled fast breeder reactor (GCBR): A fast breeder reactor which is cooled by a gas, usually helium, under pressure.

Gas, manufactured: A gas obtained by destructive distillation of coal, or by the thermal decomposition of oil, or by the reaction of steam passing through a bed of heated coal or coke. Examples are coal gases, coke-oven gases, producer gas, blast-furnace gas, blue (water) gas, carbureted water gas. Btu content varies widely.

Gas, natural: A naturally occurring mixture of hydrocarbon gases found in porous geologic formations beneath the earth's surface, often in association with petroleum. The principal constituent is methane.

Gas turbine: An engine which converts chemical energy of liquid fuel into mechanical energy by combustion. Gases resulting are expanded through a turbine.

Gaseous diffusion: The principal process for enrichment of uranium; that is, for increasing the concentration of fissionable uranium-235 in a mixture of uranium isotopes to the level required to sustain fission in a nuclear reactor.

Gasification: In the most commonly used sense, gasification refers to the conversion of coal to a high-Btu synthetic natural gas under conditions of high temperatures and pressures; in a more general sense, conversion of coal into a usable gas.

Gasoline: A petroleum fraction composed primarily of small branched-chain, cyclic, and aromatic hydrocarbons.

Generator (electric): A machine which converts mechanical energy into electrical energy.

Geothermal energy: The heat energy available in the rocks, hot water, and steam in the earth's subsurface.

Geothermal steam: Steam drawn from deep within the earth. There are about 90 known places in the continental United States where geothermal steam might be harnessed for power. These are in California, Idaho, Nevada, and Oregon.

Gross National Product (GNP): The nation's total national output of goods and services at current market prices.

Half-life, radioactive: Time required for a radioactive substance to lose 50 percent of its activity by decay. Each radionuclide has a unique half-life.

Heat: A form of kinetic energy, whose effects are produced by the vibration, rotation, and general motions of molecules.

Heat exchanger: Any device that transfers heat from one fluid (liquid or gas) to another or to the environment.

Heavy water: Deuterium oxide; that is, water in which all hydrogen atoms have been replaced by deuterium atoms.

Heat pump: A refrigeration machine that is used for heating buildings rather than cooling them. Expanding refrigeration fluid removes heat from the outside air; the fluid is then compressed, and the heat resulting from compression is discharged to a heat exchanger in the surroundings to be heated.

Heat sink: The medium or location to which waste heat is discharged.

High-temperature gas-cooled reactor (HTGR): A reactor using blocks of graphite containing fissile and fertile material and cooled with helium. HTGR's are operated at a high temperature which permits conversion of heat to electricity with improved efficiency.

Hydraulic fracturing: A general term, for which there are numerous trade or service names, for the fracturing of rock in an oil or gas reservoir by pumping a fluid under high pressure into the well. The purpose is to produce artificial openings in the rock in order to increase permeability.

Hydrocarbon: A compound containing only carbon and hydrogen. The fossil fuels are predominantly hydrocarbons, with varying amounts of organic compounds of sulfur, nitrogen, and oxygen, and some inorganic materials.

Hydroelectric plant: An electric power plant in which energy of falling water is converted into electricity by turning a turbine generator.

Joule: A unit of energy or work which is equivalent to one watt-second or 0.239 calories.

Kerosene: The petroleum fraction containing hydrocarbons that are slightly heavier than those found in gasoline and naphtha.

Kilowatt (kW): 1,000 watts. A unit of power equal to 1,000 watts or to energy consumption at a rate of 1,000 joules per second. It is usually used for electrical power. An electric motor rated at one horsepower uses electrical energy at a rate of about ¾ kilowatt.

Kilowatt-hour (kWh): A unit of work or energy equal to one kilowatt in one hour. It is equivalent to 3.6 M joules.

Kinetic energy: The energy of motion; the ability of an object to do work because of its motion.

Light-water reactor (LWR): A nuclear reactor in which ordinary water or light water is the primary coolant/moderator with slightly enriched uranium fuel. There are two commercial light-water reactor types—the boiling water reactor (BWR) and the pressurized water reactor (PWR).

Lignite: A low-grade coal of a variety intermediate between peat and bituminous coal.

Liquefaction (of coal): The conversion of coal into liquid hydrocarbons and related compounds by hydrogenation.

Liquefied natural gas (LNG): Natural gas that has been liquefied by cooling to about $-140^\circ C$. In this form, it occupies a relatively small volume and can be transported economically by ocean tanker.

Liquid-metal fast breeder (LMFBR): A nuclear breeder reactor cooled by molten sodium in which fission is accompanied by fast neutron irradiation of U-238 giving rise to Pu-239, a fissile material.

Lithium: Element No. 3 (symbol, Li; atomic weight 6.94). As found in nature, lithium consists of a mixture of two stable isotopes—lithium-6 (7.5 percent) and lithium-7 (92.4 percent). Lithium-6 is of interest as a possible fuel or source thereof for the generation of power from a controlled thermonuclear reaction.

Load: The amount of power needed to be delivered at a given point on an electric system.

Load growth: The growth in energy and power demands by a utility's customers.

Lurgi Process: The chief commercially available process for coal gasification. Having originated in Germany, this process has limited application in the United States because of problems of scaling up the size of operations and characteristics of U.S. coal. The Office of Coal Research and American Gas Association are jointly funding further development.

Margin: The difference between the net system generating capability and system maximum load requirements including net schedule transfers with other systems.

Megawatt (MW): 1,000 kilowatts, 1 million watts.

Metallurgical coal: Coal with strong or moderately strong coking properties that contains no more than 8.0 percent ash and 1.25 percent sulfur, as mined or after conventional cleaning.

Methane (CH_4): The lightest in the paraffinic series of hydrocarbons. It is colorless, odorless and flammable. It forms the major portion of marsh gas and natural gas.

Middle distillate: One of the distillates obtained between kerosene and lubricating oil fractions in the refining processes. These include light fuel oils and diesel fuel.

Mine-mouth plant: A steam-electric plant or coal gasification plant built close to a coal mine. It is usually associated with delivery of output via transmission lines or pipelines over long distances, as contrasted with plants located nearer load centers and at some distance from sources of fuel supply.

Moderator: A material, for example, graphite or water, used to absorb a large portion of the energy of fission neutrons in a nuclear reactor. By slowing down neutrons to speeds at which they tend not to be absorbed by nonfissionable nuclei, the moderator prevents quenching of the chain reaction, and permits the reactor to operate at low fuel enrichments.

National Environmental Policy Act (NEPA): An act passed in 1970 requiring that the environmental impact of most large projects and programs be considered. Among its important provisions is one requiring a detailed statement of environmental impact of and alternatives to a project to be submitted to the government before the project can begin.

Natural gas: Naturally occurring mixtures of hydrocarbon gases and vapors, the more important of which are methane, ethane, propane, butane, pentane, and hexane. The energy content of natural gas is usually taken as 1032 Btu/cu ft.

Natural uranium: Uranium as found in nature, containing 0.7 percent uranium-235, 99.3 percent of uranium-238, and a trace of uranium-234. It is also called normal uranium.

Net reserves: The recoverable quantity of an energy resource that can be produced and delivered.

Neutron: An elementary particle with approximately the mass of a proton but without any electric charge. It is one of the constituents of the atomic nucleus. It is frequently released during nuclear reactions and, on entering a nucleus, can cause nuclear reactions including nuclear fission.

Nitrogen oxides (NO_x): Chemical compounds of nitrogen (N) and oxygen (O). A product of combustion of fossil fuels whose production increases with the temperature of the process. It can become an air pollutant if concentrations are excessive.

Nuclear fission: The splitting of large atomic nuclei into two or more new nuclear species, with the release of large amounts of energy.

Nuclear fuel cycle: The various steps which involve the production, processing, use, and reprocessing of nuclear fuels.

Nuclear fusion: The process by which small atomic nuclei join together with the release of large amounts of energy.

Nuclear power plant: Any device, machine, or assembly that converts fission nuclear energy into some form of useful power, such as electrical power.

Nuclear reactor: A device in which a fission chain reaction can be initiated, maintained, and controlled. Its essential component is a core with fissionable fuel. It usually has a moderator, reflector, shielding, coolant, and control mechanisms. It is the basic machine of nuclear power.

Oil shale: A sedimentary rock containing solid organic matter from which oil can be obtained when the rock is heated to a sufficiently high temperature.

Organization of Petroleum Exporting Countries (OPEC): Founded in 1960 to unify and coordinate petroleum policies of the members. The members and the date of membership are: Abu Dhabi (1967); Algeria (1969); Indonesia (1962); Iran (1960); Iraq (1960); Kuwait (1960); Libya (1962); Nigeria (1971); Qatar (1961); Saudi Arabia (1960); and Venezuela (1960). OPEC headquarters are in Vienna, Austria.

Particulate matter: Solid particles, such as the ash, which are released from combustion processes in exhaust gases at fossil-fuel plants.

Peaking capacity: That part of a system's equipment which is operated only during the hours of highest power demand.

Peaking load: The greatest amount of all of the power loads on a system, or part thereof, which has occurred at one specified period of time.

Petroleum: An oily flammable bituminous liquid that may vary from almost colorless to black, occurs in many places in the upper strata of the earth, is a complex mixture of hydrocarbons with small amounts of other substances, and is prepared for use as gasoline, naphtha, or other products by various refining processes.

Petroleum refinery: A plant that converts crude petroleum into the many petroleum fractions (asphalt, fuel oil, gasoline, etc.). Usually this conversion is accomplished by fractional distillation.

Photon: A quantum of electromagnetic energy having properties of both a wave and a particle, but without mass or electric charge.

Photovoltaic conversion: Transformation of solar radiation directly into electricity by means of a solid-state device such as the single-crystal silicon solar cell.

Plasma: An electrically neutral, partially ionized gas in which the motion of the constituent particles is dominated by electromagnetic interactions. The study of plasma motions is called magnetohydrodynamics (MHD).

Plutonium (Pu): A heavy, fissionable, radioactive, metallic element with atomic number 94. Plutonium-239 occurs in nature in trace amounts only. However, it can be produced as a byproduct of the fission reaction in a uranium-fueled nuclear reactor and can be recovered for future use.

Pollution: The accumulation of wastes or byproducts of human activity. Pollution occurs when wastes are discharged in excess of the rate at which they can be degraded, assimilated, or dispersed by natural processes. Sometimes noxious environmental effects not caused by human activity are also called pollution.

Potential energy: Energy which is not associated with motion – thus, that which is stored in chemical bonds and in water at high elevations are forms of potential energy.

Power: The rate at which work is done or energy is transformed. Power is measured in units of work per unit time; typical units are the horsepower and the watt.

Pressurized-water reactor: A power reactor in which heat is transformed from the core to a heat exchanger by water kept under high pressure to prevent it from boiling. Steam is generated in a secondary circulation system.

Primary fuel: Fuel consumed in original production of energy, as contrasted to a conversion of energy from one form to another.

Proton: A positively charged elementary particle having a mass of about 1.7×10^{-24} grams, roughly 1840 times as great as the mass of an electron. Protons are constituents of atomic nuclei and are emitted in some nuclear reactions.

Proved reserves: The estimated quantity of crude oil, natural gas, natural gas liquids, or coal, which analysis or geological and engineering data demonstrates with reasonable certainty to be recoverable from known oil, coal, or gas fields under existing economic and operating conditions.

Pumped storage: An arrangement in which water is pumped into a storage reservoir at a higher elevation when a surplus of electricity is being generated; during times of peak demand for electricity, this water is then used for the generation of electricity as in a hydroelectric power plant.

Radiation: The process by which energy in the form of electromagnetic radiation is emitted from matter. Also, the electromagnetic or particulate rays that are emitted from atoms or molecules as they undergo internal change.

Radioactivity: The spontaneous decomposition of an atom accompanied by the release of energy.

Reactor: Any device in which a chemical or nuclear reaction is sustained in a self-supporting chain.

Refinery: An industrial complex for processing crude oil by distillation and chemical reactions so as to produce a separate petroleum product. Typical crude fractions, from top to bottom or simple to complex, are: ether, methane, and ethane (the gasolines); propane and butane; kerosene, fuel oil, and lubricants; jelly paraffin, asphalt, and tar.

Reserve generating capacity: Extra capacity maintained to generate power in the event of unusually high demand or a loss or scheduled outage of regular generating capacity.

Residual fuel oil: A high-viscosity fuel oil that must be heated before it can be pumped and handled conveniently. Residual fuel oil is the petroleum fraction that is collected after all lower-boiling fractions have been distilled away. It is used primarily in industry, in large commercial buildings, and for the generation of electricity.

Secondary recovery: Oil and gas obtained by the augmentation of reservoir energy; often by the injection of air, gas, or water into a production formation.

Slow neutron: A low-energy neutron, sometimes called a thermal neutron. The energy of a slow neutron is about 0.025 electron volts, in contrast to the energy of a fast neutron, which may exceed 1,000 electron volts. Slow neutrons are very efficient in causing fission of uranium-235. Nuclear-power plants now in operation are designed to use slow neutrons to sustain the fission reaction.

Solar cell: A device which converts solar radiation to a current of electricity.

Solar constant: The average intensity of solar radiation striking the atmosphere. The solar constant is measured on a plane perpendicular to the path of the radiation. Its value is 1.36 kilowatts per square meter.

Solar furnace: An optical device with large mirrors that focuses the rays from the sun upon a small focal point to produce very high temperatures.

Solar spectrum: The total distribution of electromagnetic radiation emitted from the sun, minus those wavelengths that are absorbed by the solar atmosphere.

Stack-gas desulfurization (scrubber): Treating of stack gases to remove sulfur compounds.

Steam power plant: A plant in which the prime movers (turbines) connected to the generators are driven by steam.

Strip mining: The mining of coal by removing covering material (the overburden) and "stripping" away the entire underlying coal seam. Other forms of coal mining are underground, and auger mining, in which coal is drilled out of seams exposed along the side of a mountain.

Synfuel: A commercially-desirable fossil fuel, normally oil or gas, obtained as a result of a conversion of a fossil fuel available in a less useful form to the commercially-desirable form. For example, the conversion of coal to synthetic gas or the conversion of oil shale to petroleum.

Synthetic natural gas (SNG): A manufactured gaseous fuel generally produced from naphtha or coal. It contains 95 to 98 percent methane, and has an energy content of 980 to 1035 Btu per standard cubic foot, about the same as that of natural gas.

Tar sands: Sedimentary rocks which contain viscous, heavy petroleum that cannot be recovered by conventional methods of petroleum production.

Tertiary recovery: Use of heat and other methods other than fluid injection to augment oil recovery (presumably occurring after secondary recovery).

Thermal pollution: An increase in the temperature of water resulting from waste heat released by a thermal electric plant, for example, added to the cooling water.

Thermal power plant: Any electric power plant which operates by generating heat and converting the heat to electricity.

Thermal reactor: A nuclear reactor in which the fission process is propagated mainly by thermal neutrons, i.e., by neutrons that have been slowed down until they are in thermal equilibrium with the atoms of the moderator.

Thermodynamics, laws of: The First Law of Thermodynamics states that energy can neither be created nor destroyed. The Second Law of Thermodynamics states that when a free exchange of heat takes place between two bodies, the heat is always transferred from the warmer to the cooler body.

Thorium (Th): A naturally radioactive element with atomic number 90 and, as found in nature, an atomic weight of approximately 232. The fertile thorium-232 isotope is abundant and can be transmuted to fissionable uranium-233 by neutron irradiation and in turn used as nuclear reactor fuel.

Ton: A unit of weight equal to 2,000 pounds in the United States, Canada, and the Union of South Africa, and to 2,240 pounds in Great Britain. The American ton is often called the short ton, while the British ton is called the long ton. The metric ton, or 1,000 kilograms, equals 2,204.62 pounds.

Tritium (T): A manmade radioactive isotope of hydrogen with two neutrons and one proton in the nucleus.

Turbine: An engine, the shaft of which is rotated by a stream of water, steam, air, or fluid from a nozzle forced against blades of a wheel.

Uranium (U): A radioactive element with the atomic number 92 and, as found in natural ores, an average atomic weight of approximately 238. The two principal natural isotopes are uranium-235 (0.7 percent of natural uranium), which is fissionable (capable of being split and thereby releasing energy), and uranium-238 (99.3 percent of natural uranium), which is fertile (having the property of being convertible to a fissionable material). Natural uranium also includes a minute amount of uranium-234.

Waste heat: Heat which is at temperatures very close to the ambient and hence is not valuable for production of power and is discharged to the environment.

Wastes, radioactive: Equipment and materials, from nuclear operations, which are radioactive and for which there is no further use. Wastes are generally classified as high-level (having radioactivity concentrations of hundreds to thousands of curies per gallon or cubic foot), low-level (in the range of 1 microcurie per gallon or cubic foot), or intermediate.

Watt: A unit of power. It is the rate of energy use or conversion when one joule of energy is used or converted per second. (A joule is about 0.25 calories.)

Watt-hour: The total amount of energy used in one hour by a device that uses one watt of power for continuous operation. Electrical energy is commonly sold by the kilowatt hour (1,000 watt-hours).

Wellhead: Oil or gas brought to the surface, ready for transportation to refinery or ship or pipeline. Wellhead costs usually refer to the cost to bring the oil or gas to the surface and do not include costs of transportation, refining, distribution, or profit.

Work: The transfer of energy from one body to another; or the energy itself, in the process of transfer. Work and energy are measured in the same units.

INDEX